T0295299

POLYMER YEARBOOK – 2011. POLYMERS, COMPOSITES AND NANOCOMPOSITES

POLYMER YEARBOOK

Additional books in this series can be found on Nova's website
under the Series tab.

Additional E-books in this series can be found on Nova's website
under the E-books tab.

POLYMER SCIENCE AND TECHNOLOGY

Additional books in this series can be found on Nova's website
under the Series tab.

Additional E-books in this series can be found on Nova's website
under the E-books tab.

POLYMER YEARBOOK – 2011. POLYMERS, COMPOSITES AND NANOCOMPOSITES

G. E. ZAIKOV, C. SIRGHIE
AND
R. M. KOZLOWSKI
EDITORS

Nova Science Publishers, Inc.
New York

NOTICE TO THE READER

Library of Congress Cataloging-in-Publication Data

ISSN: 2164-7178
ISBN 978-1-61209-645-2

Published by Nova Science Publishers, Inc. †New York

CONTENTS

PREFACE

This new book examines the chemistry and physical properties of polymers and composites within the framework of chemical kinetics, chemical physics and quantum calculations. Such an approach allows for quantitative estimation and physical treatment of oligomers, polymers and composites on the basis of polymers and filled polymers. In addition, information about calculations of bond energy in cluster aqueous nanostructures, mathematical modeling dependence of thermodynamic characteristics upon spatial-energy parameter of free atoms, exchange spatial-energy interactions and cyclohexane oxidation activated by glyoxal are a few of the topics discussed in this book.

Chapter 1 - The bond energy in some cluster aqueous structures has been calculated using spatial-energy notions and compared with quantum-mechanical methods. It is demonstrated that the variety and specifics of aqueous solutions are determined by the changes in structural energy characteristics of these cluster compounds.

Chapter 2 - The dependence of some thermodynamic characteristics upon initial spatial-energy parameters of free atoms has been analyzed. The corresponding equations have been obtained, the dissociation energies of binary molecules and enthalpy of single-atom gas formation have been calculated based on them.

Chapter 3 - The notion of spatial-energy parameter (P-parameter) is introduced based on the modified Lagrangian equation for relative motion of two interacting material points, and is a complex characteristic of important atomic values responsible for interatomic interactions and having the direct connection with electron density inside an atom. Wave properties of P-parameter are found, its wave equation having a formal analogy with the equation of Ψ-function is given.

With the help of P-parameter technique numerous calculations of exchange structural interactions are done, the applicability of the model for the evaluation of the intensity of fundamental interactions is demonstrated, initial theses of quark screw model are given.

Chapter 4 - Based on the experimental data and calculated by the semiempirical method AM1 hypersurfaces of the key competition propagation reactions, it has been assumed, that prominent promoting effect of glyoxal in the cyclohexane oxidation is attributed among others to the different activity of radicals formed and the nature and structure of the interacted species as well.

Chapter 5 - Before use liquid components of rubber compounds can be concluded in capsules with a cover of colloidal silica. At obtaining these products (BKPITS-DBS and PRS-1) capsuling is conducted in a ball mill.

Chapter 6 - The article presents the results of studies of the influence of fillers on the properties of flame retardant coating based on the phosphorusborcontaining oligomer. The influence of the content of fillers on the physical and mechanical properties of the coating has been found. The dependence of the coating properties on the amount of the filler contained in the composition has been identified.

Chapter 7 - The role of the structural peculiarities of electrical conducting polymer composites (ECPC) has been considered. Different conception on the nature of the conductivity, the mechanisms of charge transfer in heterogeneous structures are presented in this review. Experimental results obtained by different scientists only partially are in concordance with existing theoretical models. It is suggested that missing of various physical and chemical factors influencing on the processes of electrical current formation in polymer composites is one of main reasons of mentioned divergence between theory and experimental results among which the rate of the values of inter and intra phase interactions in composites may be considered as very important factor. The peculiarities of dependence of the conductivity of systems with binary conducting fillers are considered in this work too.

Chapter 8 - Effect of mechanical relaxations on the relaxation phenomena of electrical conducting polymer composites (ECPC) have been investigated. It is experimentally shown that filler content significantly affects relaxation characteristics of the material. The main reason of these differences is the effect of the filler type on proceeding of relaxational processes. Rubbers with active carbon blacks display the ranges of slow and fast relaxations much more clearly, than those with low active carbon blacks even at increased concentrations in increase of the rubber elasticity modulus with the increase of the filler concentration. The effect of the filler type is displayed by formation of an inter-phase layer and is similar to the effect of the filler content. In the case of active carbon blacks, the increase of the filler-filler and polymer-filler interactions is balanced by high content of low active carbon blacks, because in both cases the modulus of the material and internal friction in relaxation processes grow.

Chapter 9 - Trivalent metal ions have a relevant impact in different kinds of industry; for example, indium(III) and europium(III) ions are of practical importance in the development of new semiconductors and luminescent probes, respectively. Transport properties of ions and salts in aqueous solutions are important physical-chemical parameters allowing a better understanding of the behaviour of these ions in solution and so, helping to better describe the mechanism of processes taking place in their presence. However, the measurement of those transport properties is complicated due to the occurrence of hydrolysis; that may justify the scarcity of, e.g., diffusion data for aqueous solutions of europium(III) and indium(III) chlorides.

In this study, mutual diffusion coefficients for aqueous solutions of $InCl_3$ and $EuCl_3$, in a concentration range 0.002 mol dm^{-3} to 0.01 mol dm^{-3}, at 298.15 K, are reported. The open-ended conductometric capillary cell was used. The results are discussed on the basis of the Onsager-Fuoss and Pikal models.

Chapter 10 - The determining factor of the reaction of 2,6-di-tert-butylphenol with alkaline metal hydroxides is temperature, depending on which two types of potassium or sodium 2,6-di-tert-butyl phenoxides are formed with different catalytic activity in the alkylation of 2,6-di-tert-butylphenol with methyl acrylate. More active forms of 2,6-But2C6H3OK or 2,6-But2C6H3ONa are synthesized at temperatures higher than 160oC represent predominantly monomers of 2,6-di-tert-butylphenoxides producing dimmers on cooling. The data of NMR 1H, electronic, and IR spectra for the corresponding forms of 2,6-But2C6H3OK and 2,6-But2C6H3ONa isolated in the individual state showed a cyclohexadienone structure. In DMSO or DMF media the dimeric forms of 2,6-di-tert-butylphenoxides react with methyl acrylate to form methyl 3-(4-hydroxy-3,5-di-tert-butylphenyl) propionate in 64—92% yield.Key words: phenols, phenoxides, 2,6-di-tert-butylphenol, methyl acrylate, Michael reaction, kinetics, dimers, sodium hydroxide, potassium hydroxide.

Chapter 11 - Alkaline hydrolysis of diethyl *N*-acetylamino(3,5-di-*tert*-butyl-4-hydroxybenzyl) malonate is accompanied by decarboxylation. The efficiency of this process depends on the temperature and ratio of the reactants. A possibility of tautomerism with migration of the proton of phenolic hydroxyl and the influence of the structure on the antioxidation properties were considered on the basis of analysis of the IR spectral data and quantum chemical (PM6) calculation of the structures. The energies of homolysis of the OH bond of phenolic hydroxyl were calculated for a series of the synthesized compounds. It is proposed to predict the antioxidation activity on the basis of these values.

Chapter 12 - The determining factor of the reaction of 2,6-di-*tert*-butylphenol with alkaline metal hydroxides is temperature, depending on which two types of potassium or sodium 2,6-di-*tert*-butyl phenoxides are formed with different catalytic activity in the alkylation of 2,6-di-*tert*-butylphenol with methyl acrylate. More active forms of 2,6-$But_2C_6H_3OK$ or 2,6-$But_2C_6H_3ONa$ are synthesized at temperatures higher than 160 °C and represent predominantly monomers of 2,6-di-*tert*-butyl phenoxides producing dimers on cooling.

Chapter 13 - The simple theoretical model of nanocomposites polymer/organoclay reinforcement was proposed. Unlike the existing micromechanical models the offered treatment takes into account real nanofiller – polymeric matrix adhesion level. It has been shown that interfacial regions are reinforcing elements togther with nanofiller. These two structural components form nanoclay "effective particle".

Chapter 14 - It has been shown, that permeability to gas coefficient reduction at layered nanofiller introduction in polyethylene is due to polymer matrix fraction decrease, which is accessible for gas transport processes. Two models (percolation and multifractal ones) are offered for this reduction quantitative description.

Chapter 15 - It has been found out that continuous models do not give adequate description of melt viscosity for particulate-filled polymer nanocomposites. The correct treatment of the mentioned viscosity can be obtained within the frameworks of viscous liquid flow fractal model. It has been shown that such approach differs principally from the used ones at microcomposites viscosity description.

Chapter 16 - The interfacial layer thickness and its elasticity modulus were determined experimentally for particulate-filled polymer nanocomposite. It has been found out, that elasticity modulus of interfacial layer exceeds in 5 times corresponding characteristic for bulk polymer matrix. It has been shown, that the theoretical calculation of interfacial layer thickness within the frameworks of fractal model corresponds well to experimental data.

Chapter 17 - It has been shown, that aggregation (tangled coils formation) of carbon nanotubes begins at their very small contents. This factor strongly reduces reinforcement degree of nanocomposites polymer/carbon nanotubes. The estimation of the main parameters, influenced on the indicated nanocomposites elasticity modulus, was fulfilled. Theoretical calculations showed high potential of nanocomposites filled with nanotubes.

Chapter 18 - For the first time there is formulated a notion of gradiently oriented state. Some regularity for formation of gradiently oriented state in polymers are established .The specificity of influence of non-homogeneous mechanical field has been demonstrated on the example of creation of GB-elements, which is the object of research of the new non-traditional direction of the gradient optics – GB-optics (Gradient Birefringence Optics). Possible spheres of use of GB-elements are polarized compensators, polarized holography and photonics, the interference polarized GB-monochromator, the luminescence analysis, etc.

Chapter 19 - Reaction of low temperature oxidations of ethylene glycol (EG) by molecular oxygen in the presence of salts of bivalent copper and alkali both in water and in waterless solutions was investigated. It was found that at low (close to room) temperatures and the process carrying out in waterless solutions the basic product of EG oxidation is formic acid. Rise the temperature from $20 - 40^{\circ}C$ to $80 - 90^{\circ}C$ or carrying out the reaction in the water-containing solutions leads to sharp change of a direction of reaction. EG in these conditions is oxidized with primary formation of glycolic acids salts. Change of a direction of reaction is connected, apparently, with decrease stability of chelate complexes of Cu^{2+} -ions with dianionic form of EG. The mechanism of glycolic acids formation includes, possibly, a stage of two-electronic reduction of O_2 in reaction of dioxygen with monoanionic forms of EG, coordinated on Cu^{2+} -centers.

Chapter 20 - It is shown, that at research such relaxation characteristics as the tangent of mechanical losses, dissipation of mechanical energy as a result of the internal friction, measured at periodic action of monoaxial compressive stress on the sample of polymer (LDPE) in the high elasticity state, occurs development relaxation processes in the sample. In this connection it is necessary to take into account the temperature-time conditions of the experiment or introduce appropriate amendments to the results obtained.

Chapter 21 - AFM the method has been used in the analytical purposes, namely, for research of possibility of formation of the stable supramolecular structures on the basis of binuclear heteroligand complexes Q = Ni2(AcO)3(acac)MP·2H2O (MP = N-methylpirrolidon-2) at the expense of H-bonding. Earlier it has been established by us that complexes Q really are the active particles, formed in the course of alkylarens oxidation to hydroperoxides with molecular oxygen, catalyzed by system {Ni(acac)2 + MP}.

Chapter 22 - 5th International Conference on "Times of Polymer (TOP) and Composites" was held on June, 20 – 23 2010 on Ischia Island in Hotel "Continental Terme" (Naples Bay), Italy. This conference was organized by Department of Aerospace and Mechanical Engineering, Second University of Naples – SUN, Department Materials and Production Engineering and University of Naples Federico II.

Prof. Domenico Acierno (Department of Materials and Production Engineering University of Naples Federico II) and Prof. Alberto D'Amore (Engineering Schools of II University of Naples – SUN Department of Aerospace and Mechanical Engineering) were the Co-Chairmen of conference.

Chapter 23 - 14th International Scientific Conference on "Polymeric Materials" was held on September, 15 – 17 2010 in Halle (Saale), Germany. This conference was organized by Martin Luther University Halle-Wittenberg and Polymer Competence Center Halle-Merseburg in cooperation with Innovation Center Polymer Technology.

Prof. Hans-Joachim Radusch (Polymer Competence Center Halle-Merseburg) was the conference chairman. World well known scientists were included in Program Committee: René Androsch, Michael Bartke, Mario Beiner, Wolfgang Grellmann, Thomas Groth, Jörg Kreßler, Goerg Michler, Wolfgang Paul, Thomas Thurn-Albrecht, Ralf Wehrspohn.

About 350 participants from 100 research centers of 20 countries (Germany, Russia, Vietnam, Japan, France, Georgia, Czech Republic, Romania, Uzbekistan, Iran, Poland, Austria, USA, Hungary, South Africa, Algeria, South Korea, Ukraine) took part in this conference.

Chapter 24 - International symposium "Research and Education in Innovation Era" was held on the period November, 10 – 12 2010 in Arad (Romania) on the base of "Aurel Valicu" University (AVU). The organizers of the conference were The Ministry of Education, Research, Youth and Sport of Romania (Bucharest) and AVU (Arad).

About 200 scientists and students from 25 research centers of Romania, Poland, The Netherlands, Russia, Italy, Portugal, Germany, UK, Croatia, Serbia, USA, Spain and Switherland took part in this symposium.

The Organizing Committee included 34 world well-known scientists from these countries. The program of the conference included plenary lectures, parallel sessions and poster session. Rector of AVU Prof. Lizica Mihut took part in opening ceremony of the conference. Two plenary lectures were done by Prof. Rodica Zafiu, University of Bucharest ("Present-Day Tendencies in the Romanian Language") and Dr. Patricia Davies, The European Association for University Lifelong Learning, Director of European Dolceta Project ("Dolceta European Project - education for responsible and sustainable consumers").

In: Polymer Yearbook – 2011.
Editors: G. Zaikov, C. Sirghie et al. pp. 1-6

ISBN 978-1-61209-645-2
© 2011 Nova Science Publishers, Inc.

Chapter 1

CALCULATIONS OF BOND ENERGY IN CLUSTER AQUEOUS NANOSTRUCTURES

G. A. Korablev, N. V. Khokhriakov*and G. E. Zaikov***

* – Izhevsk State Agricultural Academy[1]
Izhevsk, 426000, Russia
** – Emmanuel Institute of Biochemical Physics[2]
Moscow, 119991, Russia

ABSTRACT

The bond energy in some cluster aqueous structures has been calculated using spatial-energy notions and compared with quantum-mechanical methods. It is demonstrated that the variety and specifics of aqueous solutions are determined by the changes in structural energy characteristics of these cluster compounds.

Keywords: bond, energy, cluster, nanostructures, calculations.

Based on modified Lagrangian equation for relative movement of two interacting material points the notion of spatial-energy parameter (P-parameter), which is a complex characteristic of important atomic values responsible for interatomic interactions and directly connected with electron density in atom [1].

The value of relative difference of P–parameters of interacting atoms – components of α -coefficient of structural interactions was used as the main quantitative characteristic of structural interactions in condensed media:

---

[1] Izhevsk State Agricultural Academy, Studencheskaya St. 11, Izhevsk, 426000, Russia.
[2] Emmanuel Institute of Biochemical Physics, Kosygina 4, Moscow, 119991, Russia.

$$\alpha = \frac{\dfrac{P_1 - P_2}{(P_1 + P_2)/2}}{}100\%$$

(1)

The nomogram of dependence of structural interaction degree upon the coefficient α (the same for a wide range of structures) was obtained applying the reliable experimental data. This approach allowed evaluating the degree and direction of structural interactions of phase-formation, isomorphism and solubility in multiple systems, including the molecular ones. In particular, the peculiarities of cluster-formation in the system $CaSO4 - H2O$ were studied [2].

To evaluate the direction and degree of phase-formation processes the following equations were used [1]:

1)To calculate the initial values of P-parameters:

$$\frac{1}{q^2/r_i} + \frac{1}{W_i n_i} = \frac{1}{P_{\ni}} \; ; \; \frac{1}{P_0} = \frac{1}{q^2} + \frac{1}{(Wrn)_i}, \; P_{\ni} = P_0/r_i$$

(2, 3, 4)

here: Wi - orbital energy of electrons [3]; ri – orbital radius of i-orbital [4]; $q = Z^*/n^*$ - [5, 6]; ni – number of electrons of the given orbital, Z^* and n^* - nucleus effective charge and effective main quantum number. The value P_0 will be called spatial-energy parameter (SEP), and P_E - effective P-parameter.

The calculation results by equations [2,3,4] for some elements are given in Table 1, where we can see that for hydrogen atom the values of PE–parameters considerably differ at the distances of orbital (ri) and covalent radii (R). The hybridization of valent orbitals of carbon atom were evaluated as an averaged value of P-parameters of 2S2 and 2P2-orbitals.

2) To calculate the value of PS-parameter in binary and complex structures:

$$\frac{1}{P_S} = \frac{1}{N_1 P_1} + \frac{1}{N_2 P_2} + \dots\dots$$

(5)

where N – number of homogeneous atoms in each subsystem.

The results of these calculations for some systems are given in Table 2.

3) To determine the bond energy (E) in binary and more complicated structures:

$$\frac{1}{E} \approx \frac{1}{P_E} = \frac{1}{P_1(N/K)_1} + \frac{1}{P_2(N/K)_2} + \dots\dots .$$

(6)

Here (as applied to cluster systems): K1 and K2 – number of subsystems forming the cluster system; N1 and N2 – number of homogeneous clusters [7].

Thus for C60(OH)10 k1 = 60, k2 =10.

It is assumed that the stable aqueous cluster (H2O) can have the same static number of subsystems (k) as the number of subsystems in the system interacting with it [8]. For example, aqueous cluster of N (H2O)10 type interacts with fullerene [C6OH]10.

Table 1. P-parameters of atoms calculated via the bond energy of electrons

Atom	Valent electrons	W (eV)	r_i (Å)	q^2 (eVÅ)	P_0(eVÅ)	R (Å)	$P_E = P_0/R$ (eV)
H	$1S^1$	13.595	0.5295	14.394	4.7985	0.5295 0.28	9.0624 17.137
C	$2P^1$	11.792	0.596	35.395	5.8680	0.77 0.69	7.6208 8.5043
	$2P^2$	11.792	0.596	35.395	10.061	0.77	13.066
	$2S^1$	19.201	0.620	37.240	9.0209	0.77	11.715
	$2S^2$				14.524	0.77	18.862
	$2S^2+2P^2$				24.585	0.77	31.929
	$\frac{1}{2}\left(2S^2+2P^2\right)$						15.964
O	$2P^1$	17.195	0.4135	71.383	4.663	0.66	9.7979
	$2P^2$	17.195	0.4135	71.383	11.858	0.66 0.59	17.967 20.048
	$2P^4$	17.195	0.4135	71.383	20.338	0.66	30.815

Table 2. Structural P_S-parameters

Radical, molecules	P_1 (eV)	P_2 (eV)	P_3 (eV)	P_4 (eV)	P_S (eV)	Orbitals of oxygen atom
OH	17.967	17.137			8.7712	$2P^2$
OH	9.7979	9.0624			4.7080	$2P^1$
H_2O	2×17.138 2×9.0624	17.967 17.967			11.788 9.0226	$2P^2$ $2P^2$
C_2H_5OH	2×15.964	2×9.0624	9.7979	9.0624	3.7622	$2P^1$

Apparently, cluster [(C2H5OH)6 – H2O]10 can be formed similarly to cluster [C6OH)10, that assumes the structural interaction of subsystems (C2H5OH)60 – (H2O)10.

And the interaction of aqueous clusters can be considered as the interaction of subsystems $(H2O)60 - N(H2O)60$.

Based on such notions and assumptions the bond energy in corresponding systems was calculated by the equation (6), the results are given in Table 3.

The calculation data obtained by N.V. Khokhriakov based on his quantum-mechanical technique [9] are given here for comparison.

Both techniques present comparable bond energy values (eV). Transfer multiplier:

$$(1 \; \frac{kcal}{mol} = 0.04336 \text{ eV}).$$

Besides, the technique of P-parameter allows explaining why the energy value of cluster bonds of water molecule with fullerene C60(OH)10 two times exceeds the bond energy between the molecule in cluster water (Table 3).

In accordance with the nomogram the structure phase-formation can take place only with the relative difference of their P-parameters below 25%-30%, and the most stable structures are formed at $\alpha < (6\text{-}7)\%$.

Table 3. Calculations of bond energy – E (eV)

System	C_{60}	$(OH)_{10}$	$(H_2O)_{10}$		$P_E E$ (calculation)	
	P_1/κ_1	P_2/κ_2	P_3/κ_3	N_3	Equation (6)	Quantum-mechanical
$C_{60}(OH)_{10} -$ - $N(H_2O)_{10}$	15.964/60	8.7712/10	11.788/10	1	0.174	0.176
				2	0.188	0.209
				3	0.193	0.218
				4	0.196	0.212
				5	0.197	0.204
$(H_2O)_{60} -$ - $N(H_2O)_{60}$	P_1/κ_1	P_2/κ_2	N_2			
	9.0226/60	9.0226/60	1		0.0768	0.0863
			2		0.1020	0.1032
			3		0.1128	0.1101
			4		0.1203	0.1110
			5		0.1274	0.115
$(C_2H_5OH)_{60}$ – - $(H_2O)_{10}$	P_1/κ_1 3.7622/60	P_2/κ_2 9.0226/10			0.0586	0.0607
$(C_2H_5OH)_{10}$ – - $(H_2O)_{60}$	P_1/κ_1 3.7622/10	P_2/κ_2 9.0226/60			0.1074	$\approx 0,116$

Table 4 gives the values of coefficient α in systems H-C, H-OH and H-H2O, which are within $0.44 - 7.09(\%)$.

But in the system H-C for carbon and hydrogen atoms the interactions at the distances of covalent radii were considered, but for other systems – at the distances of orbital radius.

Thus, the interaction in the system H-C at the distances of covalent radius plays a role of fermentative action, which results in the transition of dimensional characteristics of water molecules from orbital radius to the covalent one, i.e. to the formation of the system C60 (OH)10 – N(H2O)10 with bond energy between the main components two times exceeding the one between water molecules themselves.

The broad possibilities of aqueous clusters in changing their spatial-energy characteristics apparently explain all other water properties with its different names: mineral, holly, live, spring, radioactive, etc.

Table 4. Spatial-energy interactions in the system H-R, where R= C, (OH), H₂O

System	$P_1(eV)$	$P_2(eV)$	$\alpha = \dfrac{\Delta P}{<P>} 100\%$	Type of spatial bond
H-C	17.137	15.964	7.09	Covalent
H-OH	9.0624	8.7712	3.27	Orbital
H-H₂O	9.0624	9.0226	0.44	Orbital

CONCLUSIONS

1. Structural interactions in the bond H-C at the distances of covalent radius play the role of fermentative action, which results in the transition of dimensional characteristics of water molecules from orbital radius to covalent one, i.e. to the system: $C_{60}(OH)_{10}$ - $N(H_2O)_{10}$.
2. Broad possibilities of aqueous clusters in the change of their spatial-energy characteristics apparently explain all other unique properties of water with different names: mineral, holly, live, spring, radioactive, etc.

REFERENCES

[1] Korablev G.A. Spatial-Energy Principles of Complex Structures Formation, Leiden, the Netherlands, Brill Academic Publishers and VSP, 2005, 426 pages (Monograph).
[2] Korablev G.A., Yakovlev G.I., Kodolov V.I. Some peculiarities of cluster-formation in the system CaSO₄-H₂O. Chemical physics and mesoscopy, vol. 4, №2, 2002, p.188-196.
[3] Fischer C.F. Average-Energy of Configuration Hartree-Fock Results for the Atoms Helium to Radon.//Atomic Data,-1972, № 4, p. 301-399.
[4] Waber J.T., Cromer D.T. Orbital Radii of Atoms and Ions.//J. Chem. Phys -1965, -V 42, -№12, p. 4116-4123.
[5] Clementi E., Raimondi D.L. Atomic Screening constants from S.C.F. Functions, 1.//J.Chem. Phys.1963, v.38, №11, p. 2686-2689.

[6]   Clementi E., Raimondi D.L. Atomic Screening constants from S.C.F. Functions, 2.//*J.Chem. Phys.*-1967, v.47, №4, p. 1300-1307.

[7]   Korablev G.A., Zaikov G.E. Energy of chemical bond and spatial-energy principles of hybridization of atom orbitals.//*J. Applied Polymer Science*. USA, 2006,V.101, n.3, p.2101-2107.

[8]   Hodges M.P., Wales D.J. Glolal minima of protonated Water clusters. *Chemlocal Physies Letters*, 324, (2000), p.279-288.

[9]   Khokhriakov N.V., Melchor Ferrers. Electron properties of contacts in ideal carbon nanotubes. // *Chemical physics and mesoscopy*. Vol. 4, №2, 2002, p. 261-263.

In: Polymer Yearbook – 2011.
Editors: G. Zaikov, C. Sirghie et al. pp. 7-24

ISBN 978-1-61209-645-2
© 2011 Nova Science Publishers, Inc.

Chapter 2

MATHEMATICAL MODELLING DEPENDENCE OF THERMODYNAMIC CHARACTERISTICS UPON SPATIAL-ENERGY PARAMETER OF FREE ATOMS

G. A. Korablev[1], N. G. Korableva*and G. E. Zaikov[2]***

*Izhevsk State Agricultural Academy, Izhevsk 426000, Russia
**N.M. Emanuel Institute of Biochemical Physics, RAS,
Moscow 119991, Russia

ABSTRACT

The dependence of some thermodynamic characteristics upon initial spatial-energy parameters of free atoms has been analyzed. The corresponding equations have been obtained, the dissociation energies of binary molecules and enthalpy of single-atom gas formation have been calculated based on them.

Keywords: Spatial-energy parameter, formation enthalpy, sublimation, dissociation energy.

INTRODUCTION

Thermodynamic parameters (enthalpy, entropy, thermodynamic potential) allows describing various physical, chemical and other processes and explicitly assess the possibility of their flow without vividly using physical models. The application of reliable values of formation enthalpies is required for searching new perspective materials and compounds,

---

[1] Korablev Grigory Andreevich, Doctor of Chemical Science, Professor, Head of Department of Physics at Izhevsk State Agricultural Academy. Izhevsk 426052, 30 let Pobedy St., 98-14, tel.: +7(3412) 591946, e-mail: biakaa@mail.ru, korablev@udm.net.
[2] Zaikov Gennady Efremovich, Doctor of Chemical Science, Professor of N.M. Emanuel Institute of Biochemical Physics, RAS, Russia, Moscow 119991, 4 Kosygina St., tel.: +7(495)9397320, e-mail: chembio@chph.ras.ru.

assessment of molecule kinetic properties, analysis mechanisms of rocket propellant combustion, etc. [1,2,3].

The establishment of connection between the structure and thermochemical parameters, as well as kinetic and thermochemical characteristics of interacting systems is of great importance [4,5]. But the analysis of dependences between main parameters of chemical thermodynamics and spatial-energy characteristics of free atoms is still topical.

For this purpose the methodology of spatial-energy parameter (P-parameter) has been used in this research [6].

RESEARCH TECHNIQUE

The comparison of multiple regularities of physical and chemical processes allows assuming that in many cases the principle of adding reciprocals of volume energies or kinetic parameters of interacting structures is fulfilled.

Some examples are: ambipolar diffusion, cumulative rate of topochemical reaction, change in the light velocity when moving from vacuum into the given medium, resultant constant of chemical reaction rate (initial product – intermediary activated complex – final product).

Lagrangian equation for the relative motion of isolated system of two interacting material points with masses m_1 and m_2 in the coordinate x with acceleration a can be as follows:

$$\frac{1}{1/(m_1 a\Delta x) + 1/(m_2 a\Delta x)} \approx -\Delta U \qquad \text{or} \qquad \frac{1}{\Delta U} \approx \frac{1}{\Delta U_1} + \frac{1}{\Delta U_2} \qquad (1)$$

where ΔU_1 and ΔU_2 – potential energies of material points on elemental interaction section, ΔU – resultant (mutual) potential energy of these interactions.

The atom system is formed by oppositely charged masses of nucleus and electrons. In this system the energy characteristics of the subsystems are: orbital energy of electrons and effective energy of nucleus considering screening effects (by Clementi). Either the bond energy of electrons (W) or the ionization energy of an atom (E_i) can be used as the orbital energy.

Therefore, assuming that the resultant interaction energy in the system orbital-nucleus (responsible for interatomic interactions) can be calculated following the principle of adding reciprocals of some initial energy components, the introduction of P-parameter [6,7] as averaged energy characteristic of valent orbitals based on the following equations has been substantiated:

$$\frac{1}{q^2/r_i} + \frac{1}{W_i n_i} = \frac{1}{P_э} \qquad (2)$$

$$P_E = \frac{P_0}{r_i} \qquad (3)$$

$$\frac{1}{P_0} = \frac{1}{q^2} + \frac{1}{(wrn)_i} \qquad (4)$$

$$q = \frac{Z^*}{n^*} \qquad (5)$$

where: W_i – bond energy of an electron [8]; n_i – number of elements of the given orbital; r_i – orbital radius of i–orbital [9]; Z^* and n^* – effective charge of a nucleus and effective main quantum number [10,11].

The value of P_0 will be called spatial-energy parameter, and the value of P_E – effective P-parameter. The effective P_E-parameter has a physical sense of some averaged energy of valent electrons inside the atom and is measured in energy units, e.g. in electron-volts (eV).

The calculations demonstrated that the values of P_E-parameters numerically equal (in the limits of 2%) the total energy of valent electrons (U) by statistic model of an atom. Using the known relationship between the electron density (β) and interatomic potential by statistic model of an atom, the direct dependence of P_E-parameter upon the electron density on the distance r_i from the nucleus can be obtained:

$$\beta_i^{2/3} = \frac{AP_0}{r_i} = AP_E$$

where A – constant.

Based on the equations (2-5) P_E and P_0-parameters of free atoms for the majority of elements of the periodic system have been calculated [6,7]. Some of these calculations are given in Table 1.

Modifying the rules of addition for reciprocals of energy values of subsystems with reference to complex structures, the formula for calculating P_C-parameter of a complex structure is obtained:

$$\frac{1}{P_C} = \left(\frac{1}{NP_E}\right)_1 + \left(\frac{1}{NP_E}\right)_2 + \ldots \qquad (6)$$

where N_1 and N_2 – number of homogeneous atoms in subsystems.

CALCULATION OF DISSOCIATION ENERGY OF BINARY MOLECULES VIA AVERAGE VALUES OF P_o-PARAMETERS

The application of methods of valent bond and molecular orbitals to complex structures faces significant obstacles for prognosticating energy characteristics of the bonds formed.

Based on the equation (6), the application of the formula for calculating the dissociation energy has turned out to be practically suitable [12]:

$$\frac{1}{D_0} = \frac{1}{P_C} = \frac{1}{\left(P_E \frac{N}{K}\right)_1} + \frac{1}{\left(P_E \frac{N}{K}\right)_2}$$

(7)

where N – bond order, K – maxing or hybridization coefficient that usually equals the number of valent electrons considered, and the value of PE (N/K) has a physical sense of the averaged energy of spatial-energy parameter accrued to one valent electron of the orbitals registered. For complex structures PE-parameter was being averaged by all main valent orbitals.

For binary molecules the dissociation energy (Do) corresponds to the value of chemical bond energy: Do=E.

The calculation results of dissociation energy by the equation (7), given ion Table 2, demonstrated that PC=Д0. For some molecules containing such elements as F, N and O, the values of ion radius have been applied to register the bond ionic character for calculating PE-parameter (in Table 2 marked with *). For such molecules as C2, N2, O2 the calculations have been made by multiple bonds. In other cases the average values of bond energy have been applied. The calculated data do not contradict the experimental ones [2,3].

EQUATION OF DEPENDENCE OF THERMODYNAMIC CHARACTERISTICS UPON SPATIAL-ENERGY PARAMETERS OF FREE ATOMS

The dissociation energy (E) of the molecule breakage into two parts numerically equals the difference of the heat produced during the formation of dissociation products and during the initial molecule formation:

$$E = D_0 = \left[\Delta H_0(R_1) + \Delta H_0(R_2)\right] - \Delta H_0(R_1 R_2)$$

(8)

where D_0 – dissociation energy of a molecule (rupture energy of its bond), $\Delta H_0(R_1)$ and $\Delta H_0(R_1)$ – formation enthalpy at 0°K respectively of dissociation products R_1 and R_2, $\Delta H_0(R_1 R_1)$ – initial energy formation enthalpy.

Since for binary molecules $E=D_0$, the equation (7) gives:

$$E = \frac{\left(P_E \frac{N}{K}\right)_1 \left(P_E \frac{N}{K}\right)_2}{\left(P_E \frac{N}{K}\right)_1 + \left(P_E \frac{N}{K}\right)_2}$$

(7a)

From the formulas (8) and (7a) we have:

$$\frac{\left(P_E\dfrac{N}{K}\right)_1\left(P_E\dfrac{N}{K}\right)_2}{\left(P_E\dfrac{N}{K}\right)_1+\left(P_E\dfrac{N}{K}\right)_2} = \Delta H_0(R_1)+\Delta H_0(R_2)-\Delta H_0(R_1R_2) \qquad (9)$$

This is the equation of direct dependence of thermodynamic values and initial spatial-energy characteristics of a free atom.

The relationship obtained (9) allows defining the corresponding dependences with the experimental thermodynamic characteristics of chemical reactions.

Thus in thermodynamic experimental methods for determining the formation enthalpy the following equation is used:

$$\Delta H_0 = T\left(\Delta\Phi_T^* - R\ln K_E\right)$$

where: K_E – equilibrium constants of chemical reaction; T – thermodynamic temperature of the process; $\Delta\Phi_T^*$ – change in the reaction of considered reduced thermodynamic potential; R – universal gas constant.

EVALUATION OF SINGLE-ATOM GAS FORMATION ENTHALPY (ΔH_G^0)

In the experimental evaluation method of single-atom gas formation enthalpy from solids, a known thermodynamic identity can be used:

$$\Delta H_G^0 \equiv \Delta H_{SD}^0 + \Delta H_{S,298} \qquad (10)$$

where: ΔH_G^0 – formation enthalpy of gaseous substance;

ΔH_{SD}^0 – formation enthalpy of solid in nonstandard state;

$\Delta H_{S,298}$ – sublimation enthalpy (substance transition from solid into gaseous state).

The heats of formation of chemical elements in standard state are taken as zero. The list of such elements is known, e.g. [13] and mainly incorporates elements in solid state. Thus the single-atom gas formation enthalpy in these cases equals the sublimation enthalpy and is determined by physical and chemical criteria of the process. The sublimation comes to diffusion movements of heated particles into the surface layer with further effusion (outflow). The diffusion activation energies in external and internal regions are quite different [14].

It is usually believed that sublimation flows by the particle migration from more strongly bonded state with the biggest number of neighbors to less strongly bonded and further – to the adsorbed surface layer [15]. By analogy, the number of interacting molecules in liquid decreases, when molecules move from the lower part of the surface layer to its upper part.

The diffusion activation energy is defined by the values of electron densities of migrating particle and particles surrounding it at the distance of atom-molecular interaction radius (R), i.e. such interaction is carried out via the power field of particles evaluated by the values of their P-parameters [16]. In such a model the diffusion process has basically the similar nature in any aggregate state and its energy and directedness are defined by three main factors:

1) value of P-parameters of structures;
2) number of particles;
3) radius of atom-molecular inetarction.

In advanced researches carried out for solid solutions in the frames of generalized grid model it has been found out that "effective diffusion coefficient depends on local composition, own volumes of component atoms and potentials of paired interactions" [17]. At the same time two types of diffusion are also distinguished: "normal" and "bottom-up" [18], this apparently agrees with the notion of volume (internal) and surface diffusion.

In a liquid the radius of the sphere of molecular interaction $R \approx 3r$, where r – radius of a molecule. Liquids are mainly formed by the elements of first and second periods of the system. For the second period we can write down: $R \approx 3r = (n+1)r$, where n – main quantum number. For both periods (first and second) we have $R = (\langle n \rangle + 1)r \approx 2,5r$.

Let us assume that this principle with definite approximation can be extended to various aggregate states of elements of all other periods but taking into consideration screening effects introducing the value of effective main quantum number (n^*) instead of n. These values of n^* and $n^* + 1$ taken by Slater [19] are shown in Table 3.

Thus we assume that the radius of sphere of atom-molecular interaction during the particle diffusion in the sublimation process is defined as follows:

$$R = (n^* + 1)r,$$

where r – dimensional characteristic of atom structure. Total change of R is from 3r to 5,2r.

Table 1. P-parameters of atoms calculated via bond energy of electrons

Atom	Valent electrons	W (eV)	r_i (Å)	q^2_0 (eVÅ)	P_0 (eVÅ)	R (Å)	P_0/R (eV)	$P_0/R(n*+1)$	r_i (Å)	P_0/r_i (eV)	$P_0/r_i(n*+1)$
1	2	3	4	5	6	7	8	9	10	11	12
Li	2S¹	5.3416	1.506	5.86902	3.475	1.55	2.2419	0.7473	0.68	5.1103	1.7034
Na	3S1	4.9552	1.713	10.058	4.6034	1.69	2.4357	0.60892	0.96	4.6973	1.1743
K	4S1	4.0130	2.612	10.993	4.8490	2.36	2.0547	0.4372	1.33	3.6459	0.7757
Rb	5S1	3.7511	2.287	14.309	5.3630	2.48	2.1625	0.43250	1.49	3.5993	0.71986
Cs	6S1	3.3647	2.516	16.193	5.5628	2.68	2.0757	0.3992	1.65	3.3714	0.6483
Mg	3S1 3S2	6.8859 6.8859	1.279 1.279	17.501 17.501	5.8568 8.7787	1.60 1.60	3.6616 514667	0.91544 1.3717	0.74 0.74	7.9173 11.863	1.9793 2.9658
Ca	4S1 4S2	5.3212 5.3212	1.690 1.690	17.406 17.406	5.929 6.6456	1.97 1.97	3.0096 4.902	0.64035 0.95535	1.33 1.04	8.5054	1.8097
Sr	5S1 5S2	4.8559	1.836	21.224	6.790 9.6901	2.15 2.15	2.9205 4.5070	0.56409 0.90140	1.20	8.0751	1.6150
Ba	6S1 6S2	4.2872	2.060	22.950	6.3768 9.9812	2.21 2.21	2.6854 4.5164	0.5549 0.8685	1.38	7.2328	1.3909
Sc	4S1 4S2 3d1 4S2+3d1	5.7174 5.7174 9.3532 9.3532	1.570 1.570 0.539 0.539	19.311 19.311 81.099 81.099	6.1279 9.3035 4.7463 14.050	1.64 1.64 1.64	3.7365 2.8941 8.5671	0.7950 0.6158 1.8228	0.83	16.928	3.6016

Table 1. (Continued)

Atom	Valent electrons	W (eV)	r_i (Å)	q²_0 (eVÅ)	P_0 (eVÅ)	R (Å)	P_0/R (eV)	P_0/R(n*+1)	r_i (Å)	P_0/r_i (eV)	P_0/r_i(n*+1)
1	2	3	4	5	6	7	8	9	10	11	12
Y	5S1	6.3376	1.693	22.540	6.4505 10.030	1.81	3.5638	0.71276 1.1083			
	5S2				5.6756		5.5417	0.62714			
	4d1	6.7965	0.856	229.18	15.706		3.1357	1.7354	0.97	16.192	3.2384
	4d1+5S2						8.6771				
La	6S1	4.3528	1.915	34.681	6.7203	1.87	3.5937	0.6911 1.1579			
	6S2				11.259		6.0209	0.4378			
	4f1	10.302	0.4234	17870	4.2576		2.2768	1.8922 3.0501			
	5P1	25.470	0.827	145.53	18.400		9.8396	1.5957	1.04 1.04	28.518 14.920	5.4843 2.8693
	6S2+5P1				29.659		15.860				
	6S2+4f1				15.517		8.2976				
Ti	4S1	6.0082	1.477 1.477	20.879	6.2273 9.5934	1.46 1.46 1.46	4.2653	0.9075	0.78	12.299	2.6169
	4S2	6.0082	0.496 0.469	20.879	5.556 10.558		6.5708	1.3960 0.80965			
	3d1	11.990		106.04	20.151	1.46	3.8053				
	3d2	11.990		106.04				2.9366			
	4S2+3d2						13.802	1.7172	0.64	31.486	6.6981
	4S1+3d1										
Zr	5S1	5.5414	1.593	23.926	6.5330 10.263	1.60	4.0831	0.8166 1.2829			
	5S2				6.9121 13.229		6.4146	0.8640 1.6536			
	4d1	9.1611	0.790	153.76	23.492		4.3201	2.9365	0.82	28.640	5.7298
	4d2						8.2681				
	5S2+4d2					1.6	14.683				

Table 2. dissociation energies of two-atom molecules – D0 $\left(\frac{KJ}{mol}\right)$

| Structure | First atom | | | | Second atom | | | | P_C (eV) | D_0 calcul. | D_0 experim. |
	Orbitals	N/k	P_E (eV)	$P_E \dfrac{N}{k}$	Orbitals	N/k	P_E (eV)	$P_E \dfrac{N}{k}$			
1	2	3	4	5	6	7	8	9	10	11	12
CCl	2P¹	1/1	7.6208	7.6208	3P¹	1/1	8.5461	8.5461	4.0209	388.9	393.3
CBr	2P¹	1/1	7.6208	7.6208	4P¹	1/1	8.0430	8.0430	3.9130	377.7	364
CJ	2P¹	1/1	7.6208	7.6208	5P¹	1/1	7.2545	7.2545	2.2523	217.4	209.2
CN	2P²	2/2	13.066	13.066	2S²2P³	2/5	47.413	18.965	7.7358	746.7	755.6
CN	2P²	2/2	14.581	14.581	2P³	2/3	25.127	16.751	7.796	752.5	755.6
C-O	2P²	1/2	13.066	6.533	2P²	1/2	17.967	8.984	3.782	365	356
NO	2P²	1/1	9.2839	9.2839	2P²	2/2	20.048	20.048	6.346	612.5	626.8
CH	2P²	1/2	13.066	6.533	1S¹	1/1	9.066	9.066	3.7969	366.5	333±1
OH	2P²	1/2	17.967	8.9835	1S¹	1/1	9.066	9.066	4.5118	435.5	423.7
ClF	3S²3P⁵	1/7	29.391*	4.1987	2S²2P⁵	1/7	38.202*	5.4574	2.5579	246.9	229.1
ClO	3S²3P⁵	1/7	29.391*	4.1987	2P²	2/2	8.7191*	8.7191	2.8337	273.5	264
ClO	3P¹	1/1	4.7216*	4.7216	2S²2P⁴	1/6	30.738*	5.123	2.450	237.2	264
FO	2P¹	1/1	4.9887*	4.9887	2P²	1/2	8.7191*	4.3596	2.327	224.6	219.2
NF	2P³	1/3	10.696*	3.5653	2P¹	1/7	38.202*	5.4574	2.486	239.5	298.9
NCl	2P³	1/3	22.296	7.432	3P¹	1/1	8.5461	8.5461	3.9751	383.7	384.9
H₂	1S¹	1/1	9.0624	9.0624	1S¹	1/1	9.066	9.066	4.533	437.5	432.2
Li₂	2S¹	1/1	2.2419	2.2419	2S¹	1/1	2.2419	2.2419	1.121	108.2	98.99
B₂	2P¹	1/1	5.4885	5.4885	2P¹	1/1	5.4885	5.4885	2.744	264.9	276±21
C-C	2P¹	1/1	7.6208	7.6208	2P¹	1/1	7.6208	7.6208	3.810	367.8	376.7
C=C	2P²	2/2	13.066	13.066	2P²	2/2	13.066	13.066	6.533	630.6	611
N-N	2P³	1/3	10.696*	3.5653	2P³	1/3	10.696*	3.5653	1.783	172.1	161
N=N	2S²2P³	2/5	22.745*	9.098	2S²2P³	2/5	22.745*	9.098	4.549	439	418
O-O	2P²	1/2	8.7191	4.3596	2P²	1/2	8.7191	4.3596	2.1798	210.4	213.4
O=O	2S²2P⁴	2/6	30.738*	10.246	2S²2P⁴	2/6	30.738*	10.246	5.123	494.5	498.3

Table 3. Effective Main Quantum Number

n	1	2	3	4	5	6
n^*	1	2	3	3.7	4	4.2
n^*+1	2	3	4	4.7	5	5.2

When forming single-atom gases, the sublimation process is accompanied by the rupture of paired bond of atoms of near surroundings. The averaged value of structural P_S-parameter of interacting atoms can be the assessment of formation enthalpy and numerically equals the value of P_0-parameter falling on the radius unit of atom-molecular interaction but taking into consideration the relative number of interacting particles by the equation:

$$P_S = \frac{P_0}{R}\gamma = \frac{P_0\gamma}{r(n^*+1)} \approx \Delta H_\Gamma^o$$

where γ - coefficient taking into consideration the relative number of interacting particles and equaled to (as the calculations demonstrated):

$$\gamma = \frac{N_0}{N}$$

Here N_0 – number of particles in the sphere volume of the radius R,

N – number of particles of realized interactions depending on the process type (internal or surface diffusion).

Inside the liquid below the top layer 2R thick, the resultant force of molecular interaction equals zero.

Applying the initial analogy to internal diffusion and sublimation we can consider that such equilibrium state corresponds to the equality $N_0=N$, and then $\gamma=1$.

On the top part of liquid surface layer the volume of the sphere of atom-molecular interaction and the number of particles in it is practically 2 times lower in comparison with internal layers under 2R, i.e. $\frac{N_0}{N} \approx \frac{1}{2}$ and $\gamma=\frac{1}{2}$ – for surface diffusion and sublimation.

In case of volume diffusion even a more extreme option is possible, when the number of particles of realized interactions 2 times exceeds N_0.

Thus two-valent elements magnesium and calcium form either two or even four covalent bonds (in chelate compounds): two covalent bonds and two – by donor-acceptor mechanism. For such and analogous cases with volume diffusion $\gamma=2$.

Table 4. calculation of formation enthalpy of single-atom gases – Δ H298

$$1\ eV = 96.525\ \ kJ/mol$$

Group	Atom	Orbitals	P_o $\left(eVA^o\right)$	γ	$r\left(A^o\right)$	$\Delta H = \dfrac{P_o\gamma}{r(n^*+1)}$ (eV)	$r_u\left(A^o\right)$	$\Delta H = \dfrac{P_o\gamma}{r_u(n^*+1)}$ (eV)	ΔH calculation $\left(kJ/mol\right)$	ΔH° 298 ref. data $\left(kJ/mol\right)$
1	2	3	4	5	6	7	8	9	10	11
1	Li	2S^1	3.4750	1			0.68	1.703	164.4	159.3
	Na	3S^1	4.6034	1			0.98	1.1743	113.3	107.5
1a	K	4S^1	4.8490	2	2.36	0.874			84.36	88.9
	Rb	5S^1	5.3630	2	2.48	0.865			83.49	80.9
	Cs	6S^1	5.5628	2	2.68	0.7984			77.07	77
	Be	2S^2	7.512	½			0.34	3.682	355.4	326.4
	Mg	3S^2	8.7787	½			0.74	1.483	143.1	147.1
2a	Ca	4S^2	8.8456	1			1.04	1.8097	174.7	177.8
	Sr	5S^2	9.6901	1			1.20	1.6150	155.9	160.7
	Ba	6S^2	9.9812	2	2.21	1.737			167.7	179.1
3a	Sc	4S^2 3d^1	14.050	1			0.74	4.040	389.96	379.1
	Y	5S^1 4d^2	17.527	2	1.62	4.342			419.2	423
	La	6S^2 5p^1 6S^2 4f^1	29.659 15.517	2 2	1.87 1.87	6.1002 3.1915			<448.4>	429.7

Table 4. (Continued)

1	2	3	4	5	6	7	8	9	10	11
4a	Ti	$4S^1 3d^1$	11.7853	2	1.46	3.4344			<449.2>	468.6
		$4S^2 3d^2$	20.151	2	1.46	5.8732				
	Ti	$4S^1 3d^1$	11.7853	1			0.78	3.214	<478.4>	468.6
		$4S^2 3d^2$	20.151	1			0.64	6.6981		
	Zr	$5S^2 4d^2$	23.492	2	1.60	5.873			566.9	600
	Hf	$6S^2 5d^2$	24.498	2	$r_K = 1.44$	6.543			631.6	620.1
	V	$4S^2 3d^1$	15.776	1			0.67	5.6010	<512.5>	514.6
		$4S^1 3d^2$	17.665	1			0.67	5.6090		
	Nb $(5S^2 4d^3)$	$5S^2 4d^2$	25.577	1			0.67	7.6349	736.9	722.6
5a	Nb $(5S^1 4d^4)$	$5S^2 4d^2$	20.805	1			0.767	5.425	<709.4>	722.6
		$5S^1 4d^4$	30.607	1			0.65	9.2748		
	Nb $(3, 4, 5)$			1					<723.2>	722.6
	Ta	$6S^2 5d^2$	26.314	1			0.737	6.8662	<791.5>	786.1
		$6S^1 5d^3$	32.722	1			(0.66)	9.5344		
	Cr	$4S^2$	10.535	1			0.83	2.7006	<406.1>	397.5
		$4S^1 3d^2$	17.168	1			0.64	5.7074		
6a	Mo (2) $(5S^2 4d^4)$	$5S^1 4d^1$	19.574	1			0.737	5.312	<658>	656.5

1	2	3	4	5	6	7	8	9	10	11
	Mo (4) (5S¹ 4d⁴)	5S¹ 4d⁴	28.293	1			0.68	8.3215		
	W(4)	6S² 6S¹	27.879	2	1.40	3.8295			<831.5>	856.9
	W(5)	6S² 5d³	34.828		1.40	4.7841				
7a	Mn (2)	4S¹ 3d¹	12.924	1			0.91	3.0217	291.7	284.5
	Tc (3)	5S¹ 4d²	23.866	2	1.36	7.0194			674.1	657
	Re (4)	6S² 5d²	29.806	1			0.72	7.961	768.4	775.7
	Re (6)	6S² 5d⁴	44.519	½			0.52	8.232	794.6	
	Re (4,6)								<781>	
	Fe (2)	4S¹ 3d¹	12.717	1			0.80	3.3822	<418.6>	417.1
	Fe (3)	4S² 3d¹	16.664	1			0.67	5.2918		
	Co (2)	4S¹ 3d¹	12.707	1			0.78	3.4662	<436.5>	428.4
	Co (3)	4S² 3d¹	16.680	1			0.64	5.5785		
8	Ni (2)	4S¹ 3d¹	12.705	1			0.74	3.6530	<465.5>	429.3
	Ni (3)	4S¹ 3d²	16.897	1			0.60	5.992		
	Ni (2)	4S¹ 3d¹	12.705	2	1.24	4.360	0.68		420.8	429.3
	Ru (3)	5S¹ 4d²	23.636	1			0.68	6.982	671	656.8
	Rh (3)	5S² 4d¹	21.114	1			0.75	5.6304	543.5	556.5
	Pd (2)	5S²	12.057	1			0.64	3.7681	363.7	372.3
	Os (3)	5d³	25.986	2	$r_k = 1.26$	7.922			766	790.2

Table 4. (Continued)

1	2	3	4	5	6	7	8	9	10	11
8	Ir (2)	6S^1 5d^1	17.631	2	1.35	5.023			<665.8>	669.5
	Ir (4)	6S^2 5d^2	30.790	2	1.35	8.772				
	Pt (2)	6S^1 5d^1	17.381	1			0.6	5.971	576.4	565.7
	Cu (2)	4S^1 3d^1	13.242	1			0.80	3.5218	339.9	337.6
16	Ag	4d^1	9.7843	2	r$_k$ = 1.25	3.131			<282.3>	284.9
	Ag (1)	4d^1	9.7843	2	1.44	2.718				
	Au (3)	6S^1 5d^2	27.536	1	1.44	3.6774			355.0	368.8
	Zn (1)	3d^1	6.1153	1	1.39	0.9361			<127.1>	130.5
	Zn (2)	4S^2	11.085	1	1.39	1.6968				
26	Cd (2)	5S^2	11.839	½			0.99	1.1959	115.4	111.8
	Cd (2)	5S^1 4d^1	17.145	½	1.56	1.0991			106.1	111.8
	Cd (1, 2)								<110.8>	111.8
	Hg (1)	5d^1	11.266	½	1.60	0.6771			65.35	61.40
36	B	2S^2 2p^1	16.086	1	0.91	5.892			568.7	561
	Al	3S^2 3p^1	18.093	1	1.43 r$_k$ = 1.25	3.163 3.619			<327.3>	329.3

1	2	3	4	5	6	7	8	9	10	11
36	Ga	$4S^2\,4p^1$	20.760	1	1.39	3.178			<284.8>	273.0
	Ga		8.8961	2	1.39	2.723				
	In	$5S^2\,5p^1$	21.841	1	1.66	2.6314			<241.5>	238.1
		$5S^2\,5p^1$	21.841	½			0.92	2.374		
	Tl	$6S^2\,6p^1$	22.012	½			1.05	2.015	194.5	181.0
46	C	$2p^2$	10.061	1	0.77	4.3554			<723.9>	716.7
		$2S^2\,2p^2$	24.585	1	0.77	10.643				
	Si	$3p^2$	10.876	2	1.17	4.648			448.6	452
	Ge	$4p^2$	12.072	2	1.39	3.3656			<378.3>	376.6
		$4p^2$	12.072	2	1.24	4.1428				
	Ge	$4p^2$	12.072	1			0.65	3.9516	381.43	376.6
	Sn	$5p^2$	13.009	2	1.58	3.2934			317.9	302.1
	Pb	$6S^2\,6p^2$	32.526	1	1.50	2.0952			202.2	195.1
	Pb	$6p^2$	13.460				1.26	2.0543	198.3	195.1
56	N	$2p_r^{\,5}$	21.966				1.48	4.947	477.5	472.7
	P	$3p^1$	7.7864	2	$r_K = 1.16$	3.3562			323.9	316.3
	As (5)	$4S^2\,4p^3$	40.232	½	1.40	3.057			295.1	301.8
	As (3)	$4p^3$	18.645	1	$r_K = 1.21$	3.2785			316.5	
	As (3 ,5)								<305.8>	

Table 4. (Continued)

1	2	3	4	5	6	7	8	9	10	11
56	Sb	$5p^3$	20.509	1	$r_K = 1.39$	2.9509			<265.2>	268.2
		$5p^3$	20.509	1	1.61	2.5477				
	Bi	$6p^3$	21.919	1	1.82	2.3160			<207.3>	209.2
	Bi	$6p^3$	21.919	1			2.13	1.9790		
66	O (2)	$2p^2$	11.858	1			1.40	2.823	<253.1>	249.2
	O (4)	$2p^4$	20.338	½			1.40	2.421		
	S (2)	$3S^1 3p^1$	21.673	2	$r_K = 0.94$	2.882			<280.8>	277.0
	S (4)	$3p^4$	21.375	1			1.82	2.9360		
	Se (4)	$4p^4$	24.213	½	$r_K = .17$	2.2032			<226.7>	223.4
	Se (2)	$4S^1 4p^1$	23.283	1			1.98	2.495		
	Te	$5S^1 5p^1$	23.882	1			2.11	2.264	218.5	215.6
	Po	$6S^1 6p^1$	23.664	½	$r_K = 1.50$	1.516			146.4	146
76	F	$2p^1$	6.635	½			1.33	0.83145	80.3	79.5
	Cl	$3p^1$	8.5461	1			1.81	1.2804	123.6	121.3
	Br	$4p^1$	9.3068	1			1.79	1.1062	106.8	111.8
	I	$5p^1$	9.9812	½			$r^{5+} = 0.94$	1.062	102.5	106.8

CALCULATIONS AND COMPARISONS

Based on such initial statements and assumptions the value of P_S-parameter was calculated (and ΔH^0_r) by the equation (12) for the majority of elements in periodic system – Table 4. The values of covalent, atom and ion radii are basically taken by Belov-Bokiy and partially by Batsanov [20].

The structural coefficient γ was taken as equaled to 2 for covalent spatial-energy bonds at the distances of only atom or covalent radii (surface diffusion). Coefficient γ was taken as equaled to 1 with ion spatial-energy bonds of elements from subgroups «a» and group 8 in periodic system (internal diffusion). In all other cases this coefficient equals 1 or 1/2 (internal or volume diffusion).

In some cases the values of ΔH^0_r of the given element was found as average value by its two possible valent states (marked with <...>). The calculations are carried out practically for all elements of six periods irrespectively of their aggregate state that definitely could give higher accuracy than reference data.

Experimental and reference data [2,3] have a relative error within 0.5 – 1.5 (%), but in the given calculations such average error was about 5 %. Probably the search for a more rational technique of registering screening effects for clarifying the effective main quantum number can eventually bring more reliable results for evaluating ΔH^0_r.

CONCLUSIONS

The dependences of dissociation energy of binary molecules and formation enthalpies of single-atom gases upon initial spatial-energy characteristics of free atoms have been determined.

The corresponding equations, the calculations on which basically agree with the experimental and reference data, have been obtained.

REFERENCES

[1] Lebedev Yu.A., Miroshnichenko E.A. Thermal chemistry of steam-formation of organic substances. – M.: "Nauka", 1981, 216 p.

[2] Properties of inorganic compounds. Reference-book/ / Efimov A.I. et al. – L.: "Khimiya", 1983, 392 p.

[3] Rupture energy of chemical bonds. Ionization potentials and affinity to an electron. Reference-book/ / Kondratyev V.I. et al. – M.: "Nauka'', 1974, 351p.

[4] Benson S. Thermochemical kinetics. – M.: Mir, 1971, 308 p.

[5] Berlin A.A., Wolfson S.A., Enikolopyan N.S. Kinetics of polymerization processes. – M.: "Khimiya", 1978, 319 p.

[6] Korablev G.A. Spatial-Energy Principles of Complex Structures Formation, Netherlands, Brill Academic Publishers and VSP, 2005,426p. (Monograph).

[7] Korablev G.A. Spatial-energy criteria of phase-formation processes. – Publishing house "Udmurt university", Izhevsk, 2008, 494 p.

[8]   Fischer C.F. Average-Energy of Configuration Hartree-Fock Results for the Atoms Helium to RadonV//Atomic Data,-1972, -№ 4, -p. 301-399.
[9]   Waber J.T., Cromer D.T. Orbital Radii of Atoms and Ions//J. Chem. Phys -1965, -V 42, -№12, -p. 4116-4123.
[10]   Clementi E., Raimondi D.L. Atomic Screening constants from S.C.F. Functions, 1.//J.Chem. Phys.-1963, -v.38, -№11, -p. 2686-2689
[11]   Clementi E ., Raimondi D.L. Atomik Screening Constants from S.C.F. Functions, 2.//J. Chem. Phys.- 1967,- V.47, № 4,- p. 1300 – 1307.
[12]   Korablev G.A., Zaikov G.E. Energy of chemical bond and spatial-energy principles of hybridization of atom orbitalls.//J. of Applied Polymer Science. USA, 2006., V. 101, № 3, p. 283 – 293.
[13]   Thermodynamic properties of individual substances. Reference-book of Academy of Science of the USSR, ed. Glushko V.P., M.: "Science", 1978, v.1, 496 p.
[14]   Kofstad P. Deviation from stoichio, diffusion and electrical conductivity in simple metal oxides. M.: "Mir", 1975, 398 p.
[15]   Pound G.M. – S. Phys. and Chem. Ref. Data. 1972, v.1. p. 135 – 146.
[16]   Korablev G.A., Solovyev S.D. Energy of diffusion activation in metal systems//Bulletin of ISTU, 2007, №4, p.128-132.
[17]   Zaharov M.A. Grid models of multi-component solid solutions: statistic thermodynamics and kinetics//Abstract of doctoral thesis. Novgorod State University, 2008, 36 p.
[18]   Geguzin Ya.E. Ascending diffusion and diffusion aftereffect//UFN.-1986, v.149, iss.1, p. 149-159.
[19]   Batsanov S.S., Zvyagina R.A. Overlap integrals and problem of effective charges. "Nauka", Siberian branch, Novosibirsk, 1966, 386 p.
[20]   Batsanov S.S. Structural chemistry. Facts and dependencies. – M.: MSU – 2000, 292 p.

In: Polymer Yearbook – 2011.
Editors: G. Zaikov, C. Sirghie et al. pp. 25-44

ISBN 978-1-61209-645-2
© 2011 Nova Science Publishers, Inc.

Chapter 3

EXCHANGE SPATIAL-ENERGY INTERACTIONS

*G. A. Korablev[1] and G. E. Zaikov[2]***

*Izhevsk State Agricultural Academy
Basic Research and Educational Center of Chemical Physics and Mesoscopy,
URC, UrD, RAS Russia, Izhevsk, 426000,
**N.M. Emanuel Institute of Biochemical Physics, RAS, Russia, Moscow 119991,

ABSTRACT

The notion of spatial-energy parameter (P-parameter) is introduced based on the modified Lagrangian equation for relative motion of two interacting material points, and is a complex characteristic of important atomic values responsible for interatomic interactions and having the direct connection with electron density inside an atom. Wave properties of P-parameter are found, its wave equation having a formal analogy with the equation of Ψ-function is given.

With the help of P-parameter technique numerous calculations of exchange structural interactions are done, the applicability of the model for the evaluation of the intensity of fundamental interactions is demonstrated, initial theses of quark screw model are given.

Keywords: Lagrangian equations, wave functions, spatial-energy parameter, electron density, elementary particles, quarks.

1. SPATIAL-ENERGY PARAMETER

When oppositely charged heterogeneous systems interact, a certain compensation of the volume energy, which results in the decrease in the resultant energy (e.g. during the

---

[1] Korablev Grigory Andreevich, Doctor of Chemical Science, Professor, Head of Department of Physics at Izhevsk State Agricultural Academy. Izhevsk 426052, 30 let Pobedy St., 98-14, tel.: +7(3412) 591946, e-mail: biakaa@mail.ru, korablevga@mail.ru.

[2] N.M. Emanuel Institute of Biochemical Physics, 4 Kosygina St., E-mail: chembio@sky.chph.ras.ru.

hybridization of atom orbitals), takes place [1]. But this is not the direct algebraic deduction of corresponding energies. The comparison of numerous regularities of physical and chemical processes lets us assume that in such and similar cases the principle of adding reverse values of volume energies or kinetic parameters of interacting structures is observed. For instance:

1)During the ambipolar diffusion: when joint motion of oppositely charged particles is observed in the given medium (in plasma or electrolyte), the diffusion coefficient (D) is found as follows:

$$\frac{\eta}{D} = \frac{1}{a_+} + \frac{1}{a_-}$$

where a_+ and a_- - charge mobility of both atoms, η – constant coefficient.

2) Total velocity of topochemical reaction (v) between the solid and gas is found as follows:

$$\frac{1}{v} = \frac{1}{v_1} + \frac{1}{v_2}$$

where v_1 – reagent diffusion velocity, v_2 – velocity of reaction between the gaseous reagent and solid.

3) Change in the light velocity (Δv) when moving from the vacuum into the given medium is calculated by the principle of algebraic deduction of reverse values of the corresponding velocities:

$$\frac{1}{\Delta v} = \frac{1}{v} - \frac{1}{c},$$

where c – light velocity in vacuum.

4) Lagrangian equation for relative motion of the system of two interacting material points with masses m_1 and m_2 in coordinate x is as follows:

$$m_{np} x'' = -\frac{\partial U}{\partial x} \qquad \text{where} \qquad \frac{1}{m_r} = \frac{1}{m_1} + \frac{1}{m_2} \qquad (1),(1a)$$

here U – mutual potential energy of material points; m_r – reduced mass. At the same time $x'' = a$ (characteristic of system acceleration).

For elementary interaction areas Δx:

$$\frac{\partial U}{\partial x} \approx \frac{\Delta U}{\Delta x}$$

Then:

$$m_r a\Delta x = -\Delta U; \qquad \frac{1}{1/(a\Delta x)} \cdot \frac{1}{(1/m_1 + 1/m_2)} \approx -\Delta U \quad \text{or:}$$

$$\frac{1}{1/(m_1 a\Delta x) + 1/(m_2 a\Delta x)} \approx -\Delta U$$

Since in its physical sense the product $m_i a\Delta x$ equals the potential energy of each material point $(-\Delta U_i)$, then:

$$\frac{1}{\Delta U} \approx \frac{1}{\Delta U_1} + \frac{1}{\Delta U_2} \qquad\qquad (2)$$

Thus the resultant energy characteristic of the interaction system of two material points is found by the principle of adding the reverse values of initial energies of interacting subsystems.

Therefore assuming that the energy of atom valent orbitals (responsible for interatomic interactions) can be calculated by the principle of adding the reverse values of some initial energy components, the introduction of P-parameter as the averaged energy characteristic of valent orbitals is postulated based on the following equations:

$$\frac{1}{q^2/r_i} + \frac{1}{W_i n_i} = \frac{1}{P_E} \quad \text{or} \quad \frac{1}{P_0} = \frac{1}{q^2} + \frac{1}{(Wrn)_i} \; ; \; P_E = P_0/r_i \quad (3),(4),(5)$$

here: W_i – orbital energy of electrons; [2] r_i – orbital radius of i–orbital [3]; $q = Z^*/n^*$ [4,5], n_i – number of electrons of the given orbital, Z^* and n^* - effective charge of the nucleus and effective main quantum number, r – bond dimensional characteristics.

The value P_O will be called spatial-energy parameter (SEP), and value P_E – effective P–parameter (effective SEP). Effective SEP has a physical sense of some averaged energy of valent electrons in the atom and is measured in the energy units, e.g. in electron-volts (eV).

Values of P_0-parameter are tabulated constant values for the electrons of the atom given orbital.

For the dimensionality SEP can be written down as follows:

$$[P_0] = \left[q^2\right] = [E] \cdot [r] = [h] \cdot [\upsilon] = \frac{kgm^3}{s^2} = J\,m$$

where [E], [h] and [υ] – dimensionalities of energy, Plank's constant and velocity.

The introduction of P-parameter should be considered as further development of quasi-classic notions using quantum-mechanical data on the atom structure to obtain the criteria of phase-formation energy conditions. At the same time, for similarly charged systems (e.g.

orbitals in the given atom) and homogeneous systems the principle of algebraic addition of these parameters will be preserved:

$$\sum P_E = \sum (P_0/r_i); \qquad \sum P_E = \frac{\sum P_0}{r} \qquad (6),(7)$$

or: $\sum P_0 = P_0' + P_0'' + P_0''' + \dots; \qquad r\sum P_E = \sum P_0 \qquad (8),(9)$

Here P-parameters are summed by all atom valent orbitals.

To calculate the values of P_E-parameter at the given distance from the nucleus either atomic radius (R) or ionic radius (r_i) can be used instead of r depending on the bond type.

Applying the equation (8) to hydrogen atom we can write down the following:

$$K(\frac{e}{n})_1^2 = K(\frac{e}{n})_2^2 + mc^2 \lambda \qquad (10)$$

where: e – elementary charge, n_1 and n_2 – main quantum numbers, m – electron mass, c – electromagnetic wave velocity, λ - wave length, K - constant.

Using the known correlations $\nu = C/\lambda$ and λ = h/mc (where h – Plank's constant, ν – wave frequency) from the formula (10), the equation of spectral regularities in hydrogen atom can be obtained, in which $2\pi^2 e^2 /hC = K$.

EFFECTIVE ENERGY OF VALENT ELECTRONS IN ATOM AND ITS COMPARISON WITH STATISTIC MODEL

The modified Thomas-Fermi equation converted to a simple form by introducing dimensionless variables [6] is as follows:

$$U = e(V_i - V_0 + \tau_0^2) \qquad (11)$$

where: V_o – countdown potential; e – elementary charge; τ_0 - exchange and correlation corrections; V_i – interatomic potential at the distance r_i from the nucleus, U – total energy of valent electrons.

For the 21st element the comparisons of the given value U with the values of Pe-parameter are partially given in Table 1.

As it is seen form the Table 1 the parameter values of U and Pe are practically the same (in most cases with the deviation not exceeding 1-2%) *without any transition coefficients.* Multiple corrections introduced into the statistic model are compensated with the application of simple rules of adding reverse values of energy parameters, and SEP quite precisely conveys the known solutions of Thomas-Fermi equation for interatomic potential of atoms at the distance r_i from the nucleus. Namely the following equality takes place:

$$U = P_E = e \ (V_i - V_0 + \tau^2_0) \qquad (12)$$

Using the known correlation [6] between the electron density (β_i) and interatomic potential (V_i) we have:

$$\beta^{2/3}_i \approx (3e/5) \cdot (V_i - V_0); \ \beta^{2/3}_i \approx Ae \cdot (V_i - V_0 + \tau^2_0) = [Ae \cdot r_i \cdot (V_i - V_0 + \tau^2_0)]/r_i \qquad (13)$$

where A - constant. According to the formulas (12 and 13) we have the following correlation:

$$\beta^{2/3}_i = A \ P_0/r_i \qquad (14)$$

setting the connection between P_0-parameter and electron density in the atom at the distance r_i from the nucleus.

Since the value $e(V_i - V_0 + \tau^2_0)$ in Thomas-Fermi model there is a function of charge density, P_0-parameter is a direct characteristic of electron charge density in atom.

Table 1. Comparison of total energy of valent electrons in atom calculated in Thomas-Fermi statistic atom model (*U*) and with the help of approximation

Atom	Valent electrons	r_i (Å)	X	$\varphi(X)$	U (eV)	W_i(eV)	n	q^2(eV Å)	P_E(eV)
1	2	3	4	5	6	7	8	9	10
Ar	$3P^4$	0.639	3.548	0.09-0.084	35.36-33.02	12+	4	73.196	33.45
	$3S^2$	0.607	3.268	0.122-0.105	47.81 44.81	34.8 (t) 29.0	2 2	96.107 96.107	48.44 42.45
	$2P^4$	0.146	0.785	0.47	834.25	246	4	706.3	817.12
V	$4S^2$	1.401	8.508 8.23	0.0325 0.0345	7.680 8.151	7.5	2	22.33	7.730
Cr	$4S^2$	1.453	8.95 8.70	0.0295 0.0313	7.013 7.440	7	2	23.712	7.754
Mn	$4S^2$	1.278	7.76	0.0256	10.89	6.6 (t) 7.5	2 2	25.12 25.12	7.895 10.87
Fe	$4S^2$	1.227	7.562	0.0282	8.598	8.00 7.20 (t)	2 2	26.57 26.57	9.201 8.647
Co	$4S^2$	1.181	7.565 7.378	0.02813 0.03075	9.255 10.127	8 7.5 (t)	2 2	27.98 27.98	10.062 9.187
Ni	$4S^2$	1.139	7.2102	0.02596	9.183	9 7.7 (t)	2 2	29.348 29.348	10.60 9.640
Cu	$4S^2$	1.191	7.633	0.0272	9.530	7.7	2	30.717	9.639
Jn	$5S^2$	1.093	8.424 8.309	0.033 0.03415	21.30 22.03*	11.7	2	238.3	21.8
	$4d^{10}$	0.4805	3.704	0.106	155.6	20	10	258.23	145.8
Te	$5p^4$	1.063	8.654 8.256	0.0335 0.0346	23.59 24.37*	9.8	4	67.28	24.54
	$5S^2$	0.920	7.239 7.146	0.0326 0.0341	26.54 27.72*	19 17	2 2	90.577 90.537	27.41 25.24

Note: 1) Bond energies of electrons W_i are obtained: "t" – theoretically (by Hartry-Fock method), "+" – by XPS method, all the rest – by the results of optic measurements; 2) "*" – energy of valent electrons (*U*) calculated without Fermi-Amaldi amendment.

This is confirmed by an additional check of equality correctness (14) using Clementi function [7]. A good correspondence between the values β_i, calculated via the value P_0 and obtained from atomic functions (Figure 1) is observed.

3. WAVE EQUATION OF P-PARAMETER

For the characteristic of atom spatial-energy properties two types of P-parameters with simple correlation between them are introduced:

$$P_E = \frac{P_0}{R}$$

where R – atom dimension characteristic. Taking into account additional quantum characteristics of sublevels in the atom, this equation in coordinate x can be written down as follows:

$$\Delta P_E \approx \frac{\Delta P_0}{\Delta x} \qquad \text{or } \partial P_E = \frac{\partial P_0}{\partial x}$$

where the value ΔP equals the difference between P_0-parameter of i-orbital and P_{CD}-countdown parameter (parameter of basic state at the given set of quantum numbers).

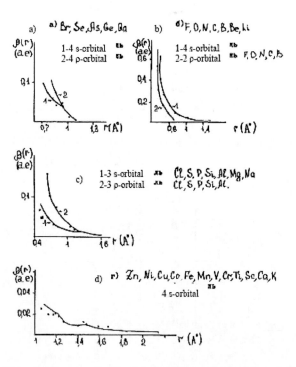

Figure 1.Electron density at the distance ri, calculated via Clementi functions (solid lines) and with the help of P-parameter (dots).

According to the established rule [8] of adding P-parameters of similarly charged or homogeneous systems for two orbitals in the given atom with different quantum characteristics and in accordance with the law of energy conservation we have:

$$\Delta P_E^{''} - \Delta P_E^{'} = P_{E,\lambda}$$

where $P_{E,\lambda}$ – spatial-energy parameter of quantum transition.

Taking as the dimension characteristic of the interaction $\Delta\lambda = \Delta x$, we have:

$$\frac{\Delta P_0^{''}}{\Delta\lambda} - \frac{\Delta P_0^{'}}{\Delta\lambda} = \frac{P_0}{\Delta\lambda} \quad \text{or:} \quad \frac{\Delta P_0^{'}}{\Delta\lambda} - \frac{\Delta P_0^{''}}{\Delta\lambda} = -\frac{P_0\lambda}{\Delta\lambda}$$

We divide termwise by $\Delta\lambda$:

$$\left(\frac{\Delta P_0^{'}}{\Delta\lambda} - \frac{\Delta P_0^{''}}{\Delta\lambda} \right) \Big/ \Delta\lambda = -\frac{P_0}{\Delta\lambda^2},$$

where:

$$\left(\frac{\Delta P_0^{'}}{\Delta\lambda} - \frac{\Delta P_0^{''}}{\Delta\lambda} \right) \Big/ \Delta\lambda \sim \frac{d^2 P_0}{d\lambda^2},$$

i.e.: $\dfrac{d^2 P_0}{d\lambda^2} + \dfrac{P_0}{\Delta\lambda^2} \approx 0$

Taking into account the interactions where $2\pi\Delta x = \Delta\lambda$ (closed oscillator), we have the following equation:

$$\frac{d^2 P_0}{dx^2} + 4\pi^2 \frac{P_0}{\Delta\lambda^2} \approx 0$$

As $\Delta\lambda = \dfrac{h}{mv}$, then:

$$\frac{d^2 P_0}{dx^2} + 4\pi^2 \frac{P_0}{h^2} m^2 v^2 \approx 0$$

or

$$\frac{d^2 P_0}{dx^2} + \frac{8\pi^2 m}{h^2} P_0 E_k = 0 \qquad (15)$$

where

$$E_k = \frac{mV^2}{2} \text{ - electron kinetic energy.}$$

Schrödinger equation for stationary state in coordinate x:

$$\frac{d^2 \psi}{dx^2} + \frac{8\pi^2 m}{h^2} \psi E_k = 0 \qquad (16)$$

Comparing the equations (15 and 16) we can see that P_0-parameter correlates numerically with the value of Ψ-function:

$$P_0 \approx \Psi$$

and in general case is proportional to it: $P_0 \sim \Psi$. Taking into account wide practical application of P-parameter methodology, we can consider this criterion the materialized analog of Ψ-function.

Since P_0-parameters like Ψ-function possess wave properties, the principles of superposition should be executed for them, thus determining the linear character of equations of adding and changing P-parameters.

4. WAVE PROPERTIES OF P-PARAMETERS AND PRINCIPLES OF THEIR ADDITION

Since P-parameter possesses wave properties (by the analogy with Ψ'-function), then the regularities of the interference of corresponding waves should be mainly executed with structural interactions.

Interference minimum, oscillation attenuation (in anti-phase) takes place if the difference in wave motion (Δ) equals the odd number of semi-waves:

$$\Delta = (2n+1)\frac{\lambda}{2} = \lambda(n + \frac{1}{2}), \quad \text{where } n = 0, 1, 2, 3, \ldots \qquad (17)$$

As applied to P-parameters this rules means that interaction minimum occurs if P-parameters of interacting structures are also "in anti-phase" – there is an interaction either between oppositely charged systems or heterogeneous atoms (for example, during the formation of valent-active radicals CH, CH_2, CH_3, NO_2 ..., etc).

In this case the summation of P-parameters takes place by the principle of adding the reverse values of P-parameters – equations (3,4).

The difference in wave motion (Δ) for P-parameters can be evaluated via their relative value ($\gamma=\dfrac{P_2}{P_1}$) or via the relative difference in P-parameters (coefficient α), which with the minimum of interactions produce an odd number:

$$\gamma=\frac{P_2}{P_1}=(n+\frac{1}{2})=\frac{3}{2};\frac{5}{2}\ \ldots$$

When n=0 (main state)

$$\frac{P_2}{P_1}=\frac{1}{2} \tag{18}$$

Let us mention that for stationary levels of one-dimensional harmonic oscillator the energy of these levels $\varepsilon=h\nu(n+\dfrac{1}{2})$, therefore in quantum oscillator, in contrast to a classical one, the minimum possible energy value does not equal zero.

In this model the interaction minimum does not produce the zero energy, corresponding to the principle of adding the reverse values of P-parameters – equations (3,4). *Interference maximum*, oscillation amplification (in phase) takes place if the difference in wave motion equals the even number of semi-waves: $\Delta=2n\dfrac{\lambda}{2}=\lambda n$

or $\Delta=\lambda(n+1)$

As applied to P-parameters the maximum amplification of interactions in the phase corresponds to the interactions of similarly charged systems or systems homogeneous in their properties and functions (for example, between the fragments and blocks of complex organic structures, such as CH_2 and NNO_2 in octogen).

Than:

$$\gamma=\frac{P_2}{P_1}=(n+1) \tag{19}$$

By the analogy, for "degenerated" systems (with similar values of functions) of two-dimensional harmonic oscillator the energy of stationary states: $\varepsilon = h\nu(n+1)$

In this model the interaction maximum corresponds to the principle of algebraic addition of P-parameters – equations (6-8). When n=0 (basic state) we have $P_2 = P_1$, or: interaction maximum of structures takes place when their P-parameters equal. This postulate can be used as [8] the main condition of isomorphic replacements.

5. STRUCTURAL EXCHANGE SPATIAL-ENERGY INTERACTIONS

In the process of solution formation and other structural interactions the single electron density should be set in the points of atom-component contact. This process is accompanied by the redistribution of electron density between the valent areas of both particles and transition of the part of electrons from some external spheres into the neighboring ones. Apparently, frame atom electrons do not take part in such exchange.

Obviously, when electron densities in free atom-components are similar, the transfer processes between boundary atoms of particles are minimal; this will be favorable for the formation of a new structure. Thus the evaluation of the degree of structural interactions in many cases means the comparative assessment of the electron density of valent electrons in free atoms (on averaged orbitals) participating in the process.

The less the difference $(P'_0/r_i - P''_0/r_i)$, the more favorable is the formation of a new structure or solid solution from the energy point.

In this regard, the maximum total solubility, evaluated via the coefficient of structural interaction α, is determined by the condition of minimum value α, which represents the relative difference of effective energies of external orbitals of interacting subsystems:

$$\alpha = \frac{P'o/r_i' - P''o/r_i''}{(P'o/r_i' + P''o/r_i'')/2}100\%$$

$$\alpha = \frac{P'_S - P''_S}{P'_S + P''_S}200\%$$

(20, 20a),

where P_S – structural parameter is found by the equation:

$$\frac{1}{P_S} = \frac{1}{N_1 P'_E} + \frac{1}{N_1 P''_E} + \dots$$

(20б),

here N_1 and N_2 – number of homogeneous atoms in subsystems.

The nomogram of the dependence of structural interaction degree По всем полученным данным была построена номограмма зависимости степени структурного взаимодействия (ρ) on the coefficient α, unified for the wide range of structures was prepared based on all the data obtained. In Figure 2 you can see such nomogram obtained using P_E-parameters calculated via the bond energy of electrons (w_i) for structural interactions of isomorphic type.

The mutual solubility of atom-components in many (over a thousand) simple and complex systems was evaluated using this technique. The calculation results are in compliance with theoretical and experimental data [8].

Isomorphism as a phenomenon is used to be considered as applicable to crystalline structures. But similar processes can obviously take place between molecular compounds, where their role and importance are not less than those of purely coulomb interactions.

In complex organic structures during the interactions the main role can be played by separate "blocks" or fragments. Therefore it is necessary to identify these fragments and evaluate their spatial-energy parameters. Based on the wave properties of P-parameter, the overall P-parameter of each fragment should be found by the principle of adding the reverse values of initial P-parameters of all atoms. The resultant P-parameter of the fragment block or all the structure is calculated by the rule of algebraic addition of P-parameters of the fragments constituting them.

The role of the fragments can be played by valent-active radicals, e.g. CH, CH_2, $(OH)^-$, NO, NO_2, $(SO_4)^{2-}$, etc. In complex structures the given carbon atom usually has two or three side bonds. During the calculations by the principle of adding the reverse values of P-parameters the priority belongs to those bonds, for which the condition of interference minimum is better performed. Therefore the fragments of the bond C-H (for CH, CH_2, CH_3 ...) are calculated first, then separately the fragments N-R, where R is the binding radical (for example – for the bond C-N).

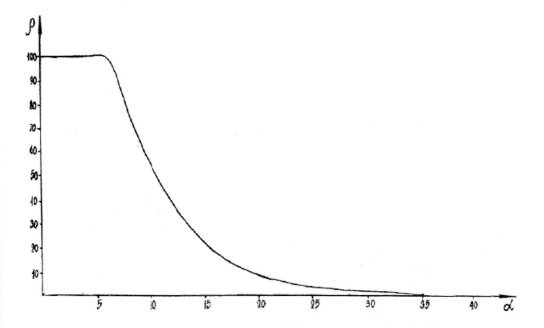

Figure 2. Dependence of the structural interaction degree (ρ) on the coefficient α.

Apparently spatial-energy exchange interactions (SEI) based on equalizing electron densities of valent orbitals of atom-components have in nature the same universal value as purely electrostatic coulomb interactions, but they supplement each other. Isomorphism known from the time of E. Mitscherlich (1820) and D.I. Mendeleev (1856) is only the particular manifestation of this general natural phenomenon. The numerical side of the evaluation of isomorphic replacements of components both in complex and simple systems rationally fit in the frameworks of P-parameter methodology. More complicated is to evaluate the degree of structural SEI for molecular, including organic structures. The technique for calculating P-parameters of molecules, structures and their fragments is successfully implemented. But such structures and their fragments are frequently not completely isomorphic with respect to each other. Nevertheless there is SEI between them, the degree of which in this case can be evaluated only semi-quantitatively or qualitatively. By the degree of isomorphic similarity all the systems can be divided into three types:

I Systems mainly isomorphic to each other – systems with approximately similar number of *dissimilar a*toms and summarily similar geometrical shapes of interacting orbitals.

II Systems with *the limited isomorphic similarity* – such systems, which:

1) either differ by the number of dissimilar atoms but have summarily similar geometrical shapes of interacting orbitals;

2) or have definite differences in geometrical shapes of orbitals but similar number of interacting dissimilar atoms.

III Systems not having isomorphic similarity – such systems, which considerably differ both by the number of dissimilar atoms and geometric shapes of their orbitals.

Then taking into account the experimental data, all types of SEI can be approximately classified as follows:

Systems I

1. $\alpha < (0\text{-}6)\%$; $\rho = 100$ %. Complete isomorphism, there is complete isomorphic replacement of atom-components;

2. $6\% < \alpha < (25\text{-}30)\%$; $\rho = 98 - (0\text{-}3)$ %.

There is either wide or limited isomorphism according to nomogram 1.

3. $\alpha > (25\text{-}30)$ %; no SEI

Systems II

1. $\alpha < (0\text{-}6)\%$;

a) There is the reconstruction of chemical bonds, can be accompanied by the formation of a new compound;

b) Breakage of chemical bonds can be accompanied by separating a fragment from the initial structure, but without attachments or replacements.

2. $6\% < \alpha < (25\text{-}30)\%$; limited internal reconstruction of chemical bonds without the formation of a new compound or replacements is possible.

3. $\alpha > (20\text{-}30)$ %; no SEI

Systems III

1. $\alpha <$ (0-6)%; a) Limited change in the type of chemical bonds of the given fragment, internal regrouping of atoms without the breakage from the main part of the molecule and without replacements;

b) Change in some dimensional characteristics of the bond is possible;

2. 6 % $< \alpha <$ (25-30)%;

Very limited internal regrouping of atoms is possible;

3. $\alpha >$ (25-30) %; no SEI.

Nomogram (Figure 2) is obtained for isomorphic interactions (systems of types I and II).

In all other cases the calculated values α and ρ refer only to the given interaction type, the nomogram of which can be clarified by reference points of etalon systems. If we take into account the universality of spatial-energy interactions in nature, this evaluation can be significant for the analysis of structural rearrangements in complex biophysical-chemical processes.

Fermentative systems contribute a lot to the correlation of structural interaction degree. In this system the ferment structure active parts (fragments, atoms, ions) have the value of P_E-parameter that is equal to P_E-parameter of the reaction final product. This means the ferment is structurally "tuned" via SEI to obtain the reaction final product, but it will not included into it due to the imperfect isomorphism of its structure (in accordance with III).

The most important characteristics of atomic-structural interactions (mutual solubility of components, energy of chemical bond, energetics of free radicals, etc) were evaluated in many systems using this technique [8-15].

6. TYPES OF FUNDAMENTAL INTERACTIONS

According to modern theories, the main types of interactions of elementary particles, their properties and specifics are mainly explained by the availability of special complex "currents" – electromagnetic, proton, lepton, etc. Based on the foregoing model of spatial-energy parameter the exchange structural interactions finally come to flowing and equalizing the electron densities of corresponding atomic-molecular components. The similar process is obviously appropriate for elementary particles as well. It can be assumed that in general case interparticle exchange interactions come to the redistribution of their energy masses – M.

The elementary electrostatic charge with the electron as a carrier is the constant of electromagnetic interaction.

Therefore for electromagnetic interaction we will calculate the system proton-electron.

For strong internucleon interaction that comes to the exchange of π-mesons, let us consider the systems nuclides-π-mesons. Since the interactions can take place with all three mesons (π^-, π^0, π^+), we take the averaged mass in the calculations (<M>=136,497 MeV/s^2).

Rated systems for strong interaction:

1) P - (π^-, π^0, π^+);
2) (P-n) - (π^-, π^0, π^+);
3) (n-P-n) - (π^-, π^0, π).

Neutrino (electron, muonic) and its antiparticles were considered as the main representatives of weak interaction.

Dimensional characteristics of elementary particles (r) were evaluated in femtometer units (1 fm = 10^{-15} m) – by the data in [16].

At the same time, the classic radius: $r_e = e^2/m_e s^2$ was used for electron, where e – elementary charge, m_e – electron mass, s – light speed in vacuum.

The fundamental Heisenberg length ($6.690 \cdot 10^{-4}$ fm) was used as the dimensional characteristic of weak interaction for neutrino [16].

The gravitational interaction was evaluated via the proton P-parameter at the distance of gravitational radius ($1.242 \cdot 10^{-39}$ fm).

In the initial equation (3) for free atom P_0-parameter is found by the principle of adding the reverse values q^2 and wr, where q – nucleus electric charge, w – bond energy of valent electron.

Modifying the equation (3), as applied to the interaction of free particles, we receive the addition of reverse values of parameters P=Mr for each particle by the equation:

$$1/ P_0 = 1/ (Mr)_1 + 1/(Mr)_2 + \dots \tag{21}$$

where M – energy mass of the particle (MeV/s²).

By the equation (21) using the initial data [16] P_0-parameters of coupled strong and electromagnetic interactions were calculated in the following systems:

nuclides-π-mesons – (P_n-parameters);

proton-electron – (P_e-parameter).

For weak and gravitational interactions only the parameters P_v=Mr and P_r=Mr were calculated, as in accordance with the equation (21) the similar nuclide parameter with greater value does not influence the calculation results.

The relative intensity of interactions (Table 2) were found by the equations for the following interactions:

1) strong $\alpha_B = < P_n > / < P_n > = P_n / P_n = 1$ (22a)

2) electromagnetic $\alpha_B = P_e / < P_n > = 1/136.983$ (22b)

3) weak $\alpha_B = P_g / < P_n >$, $\alpha_B = 2.04 \cdot 10^{-10}$; $4.2 \cdot 10^{-6}$ (22c)

4) gravitational $\alpha_B = P_\Gamma / < P_n > = 5.9 * 10^{-39}$ (22d)

In the calculations of α_B the value of P_n-parameter was multiplied by the value equaled $2\pi/3$, i.e. $<P>=(2\pi/3)P_n$. Number 3 for nuclides consisting of three different quarks is "a magic" number (see the next section for details). As it is known, number 2π has a special value in quantum mechanics and physics of elementary particles. In particular, only the value of 2π correlates theoretical and experimental data when evaluating the sections of nuclide interaction with each other [17].

As it is known [18], "very strong", "strong" and "moderately strong" nuclear interactions are distinguished. For all particles in the large group with relatively similar mass values of mass – unitary multiplets or supermultiplets – very strong interactions are the same [18]. In the frames of the given model a very strong interaction between the particles corresponds to the maximum value of P-parameter P=Mr (coupled interaction of nuclides). Taking into consideration the equality of dimensional characteristics of proton and neutron, by the equation (21) we obtain the values of P_n-parameter equaled to 401.61; 401.88 and 402.16 (MeVfm/s2) for coupled interactions p-p, p-n and n-n, respectively, thus obtaining the average value α_B =4.25. It is a very strong interaction. For eight interacting nuclides $\alpha_B \approx 1.06$ – a strong interaction.

When the number of interacting nuclides increases, α_B decreases – moderately strong interaction. Since the nuclear forces act only between neighboring nucleons, the value α_B cannot be very small.

The expression of the most intensive coupled interaction of nuclides is indirectly confirmed by the fact that the life period of double nuclear system appears to be much longer than the characteristic nuclear time [19].

Thus it is established that the intensity of fundamental interactions is evaluated via P_n-parameter calculated by the principle of adding the reverse values in the system nuclides-π-mesons. Therefore it has the direct connection with Plank's constants:

$$(2\pi/3)P_n \approx Er = 197.3 \text{ MeVfm/s}^2 \tag{23}$$

$$(2\pi/3)P_T \approx M_n \lambda_k = \text{MeVfm/s}^2 \tag{23a}$$

where E and r – Plank's energy and Plank's radius calculated via the gravitational constant; M_n, λ_k – energy mass and nuclide Compton wave-length.

In the equation (21) the exchange interactions are evaluated via the initial P-parameters of particles equaled to the product of mass by the dimensional characteristic: P=Mr.

Since these P-parameters can refer to the particles characterizing fundamental interactions, their direct correlation defines the process intensity degree (α_B):

$$\alpha_B = \frac{P_i}{P_n} = \frac{(Mr)_i}{(Mr)_n} \tag{24}$$

The calculations by the equation (24) using the known Plank's values and techniques are given in Table 3. As before, the energy and dimensional characteristics are taken from [16].

The results obtained are in accordance with theoretical and experimental data, e.g. [20, 21]

7. ON QUARK SCREW MODEL

Let us proceed from the following theses and assumptions:

1) By their structural composition macro- and micro world resemble a complex matreshka. One part has some similarity with the other: solar system – atom – atom nucleus – quarks.

2) All parts of this "matreshka" are structural formations.

3) Main property of all systems – motion: translatory, rotary, oscillatory.
4) Description of these motions can be done in Euclid three-dimensional space with coordinates x,y,z.
5) Exchange energy interactions of elementary particles are carried out by the redistribution of their energy mass M (MeV/s^2).
6) Based on these theses we suggest discussing the following screw model of the quark.
7) Quark structure is represented in certain case as a spherical one, but in general quark is a flattened (or elongated) ellipsoid of revolution. The revolution takes place around the axis (x) coinciding with the direction of angular speed vector, perpendicular to the direction of ellipsoid deformation.
8) Quark electric charge (q) is not fractional but integer, but redistributed in three-dimensional space with its virtual concentration in the directions of three coordinate axes: q/3.
9) Quark spherical or deformed structure has all three types of motion. Two of them – rotary and translatory are in accordance with the screw model, which beside these two motions, also performs an oscillatory motion in one of three mutually perpendicular planes: xoy, xoz, yoz (Figure 2).

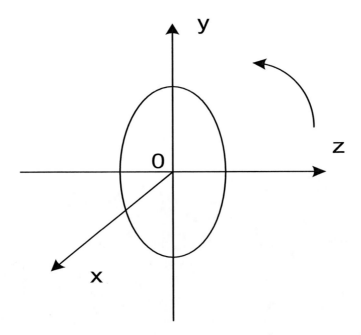

Figure 2. Structural scheme of quark in section yoz.

Table 2. Types of fundamental interactions

Interaction type		M, <M> (MeV/s²)	r (fm)	Elementary particles	M, <M> (MeV/s²)	r (fm)	P_B, P_e, P_v, P_r (MeVfm/s²)	$\dfrac{2\pi}{3}P_n$ =<P_n>	α_B , <α_B> (calculat.) – by equat. (22)	α_B (experiment)
Electromagnetic	P	938.28	0.856	e⁻	0.5110	2.8179	P_e=1.4374	-	1/136.983	1/137.04
Strong	P	938.28	0.856	π^-, π^0, π^+	136.497	0.78	P_n=94.0071	196.89	1	1
	P-n	938.92	0.856	π^-, π^0, π^+	136.497	0.78	P_n=94.015	196.90	1	1
	n-P-n	939.14	0.856	π^-, π^0, π^+	136.497	0.78	P_n=94.018	196.91	1	1
Weak				$\overline{\nu}_e$	<6·10⁻⁵	6.69·10⁻⁴	P_v=4.014·10⁻⁸		<2.04·10⁻¹⁰	10⁻¹⁰ - 10⁻¹⁴
				$\overline{\nu}_\mu$	<1.2	6.69·10⁻⁴	P_v=8.028·10⁻⁴		<4.2·10⁻⁶	10⁻⁵ - 10⁻⁶
Gravitational	P	938.28	1.242·10⁻³⁹				P_r=1.17·10⁻³⁶		5.9·10⁻³⁹	10⁻³⁸ - 10⁻³⁹

Table 3. Evaluation of the intensity of fundamental interactions using Plank's constants and parameter P=Mr

Interaction type	Particles, constants	M (MeV/s²)	r (fm)	Mr (MeVfm/s²)	α_B=Mr/(Mr)$_p$ (calculation)	α_B (experiment)
Strong	Proton	938.28	λ=0.2103	197.3	1	1
	Plank's values	1.221·10²²	1.616·10⁻²⁰	197.3		
Electromagnetic	электрон	0.5110	2.8179	1.43995	1/137.02	1/137.036
Weak	$\overline{\nu}_e$	<6·10⁻⁵	6.69·10⁻⁴	<4.014·10⁻⁸	<2.03·10⁻¹⁰	10⁻¹⁰·10⁻¹⁴
	$\overline{\nu}_\mu$	<1.2	6.69·10⁻⁴	<8.028·10⁻⁴	<4.07·10⁻⁶	10⁻⁵ - 10⁻⁶
Gravitational	Proton	938.28	1.242·10⁻³⁹ Gravitational radius	1.165·10⁻³⁶	5.91·10⁻³⁹	10⁻³⁸ - 10⁻³⁹

1) Each of these oscillation planes corresponds to the symbol of quark color, e.g. xoy – red, xoz – blue, yoz – green.

2) Screw can be "right" or "left". This directedness of screw rotation defines the sign of quark electric charge. Let us assume that the left screw corresponds to positive and right – negative quark electric charge.

3) Total number of quarks is determined by the following scheme: for each axis (x,y,z) of translator motion two screws (right and left) with three possible oscillation planes.

4) We have: $3 \cdot 2 \cdot 3 = 18$. Besides, there are 18 antiquarks with opposite characteristics of screw motions. Total: 36 types of quarks.

5) These quark numbers can be considered as realized degrees of freedom of all three motions (3 translatory + 2 rotary + 3 oscillatory).

6) Translatory motion is preferable by its direction, coinciding with the direction of angular speed vector. Such elementary particles constitute our World. The reverse direction is less preferable – this is "Antiworld".

7) Motion along axis x in the direction of the angular speed vector, perpendicular to the direction of ellipsoid deformation, is apparently less energy consumable and corresponds to the quarks U and d, forming nuclides. Such assumption is in accordance with the values of energy masses of quarks in the composition of androns: 0.33; 0.33; 0.51; 1.8; 5; (?) in GeV/s^2 for d,u,s,c,b,t – types of quarks, respectively.

The quark screw model can be proved by other calculations and comparisons.

Calculation of Energy Mass of Free Nuclide (On the Example of Neutron)

Neutron has 3 quarks d_1-u-d_2 with electric charges -1, +2, -1, distributed in three spatial directions, respectively. Quark u cements the system electrostatically. Translatory motions of the screws d_1-u-d_2 proceed along axis x, but oscillatory ones – in three different mutually perpendicular planes (Pauli principle is realized).

Apparently, in the first half of oscillation period u-quark oscillates in the phase with d_1-quark, but in the opposite phase with d_2-quark. In the second half of the period everything is vice versa. In general such interactions define the geometrical equality of directed spatial-energy vectors, thus providing the so-called quark discoloration.

The previously formulated rules of adding P-parameters spread to both types of P-parameters (P_0 and P_E). In this case, there is an addition of energy P_E-parameters, since the subsystems of interactions possess similar dimensional characteristics. As both interactions are realized inside the overall system, P_E-parameters are added algebraically, and more accurately in this case – geometrically by the following formula:

$$\frac{M}{2} = \sqrt{m_1^2 + m_2^2}$$

where, M – energy mass of free neutron, $m_1 = m_2 = 330$ MeV/s^2 masses of quarks u,d (in the composition of androns).

The calculation gives M=933.38 MeV/s². This is for strong interactions. Taking into account the role of quarks in electromagnetic interactions [21], we get the total energy mass of a free neutron: M=933.38 + 933.38/137 = 940.19 MeV/s². With the experimental value M = 939.57 MeV/s² the relative error in calculations is 0.06%.

Calculation of Bond Energy of Deuteron Via the Masses of Free Quarks

The particle deuteron is formed during the interaction of a fee proton and neutron. The bond energy is usually calculated as the difference of mass of free nucleons and mass of a free deuteron. Let us demonstrate the dependence of deuteron bond energy on the masses of free quarks. The quark masses are added algebraically in the system already formed: in proton m_1 = 5+5+7=17 MeV/s², in neutron m_2=7+7+5=19 MeV/s². As a dimensional characteristic of deuteron bond we take the distance corresponding to the maximum value of nonrectangular potential pit of nucleon interaction. By the graphs experimentally obtained we know that such distance approximately equals 1.65 fm. Exchange energy interactions of proton and neutron heterogeneous systems are evaluated based on the equation (21). Then we have:

$$1/ (M_C 1.65 K) = 1/(17 \cdot 0.856) + 1/(19 \cdot 0.856),$$

where K=2π/3. Based on the calculations we have M_C=2.228 MeV/s², this is practically coincide with reference data (M_C=2.225 MeV/s²).

Modifying the basic theses of quark screw model, it can also be applied to other elementary particles (proton, electron, neutron, etc). For instance, an electrically neutral particle neutron can be considered as a mini-atom, the analog of hydrogen atom.

CONCLUSIONS

1) The notion of spatial-energy parameter (P-parameter) is introduced based on the simultaneous accounting of important atomic characteristics and modified Lagrangian equation.
2) Wave properties of P-parameter are found, its wave equation formally similar to the equation of ψ-function is obtained.
3) Applying the methodology of P-parameter:
   a) most important characteristics of exchange energy interactions in different systems are calculated;
   b) intensities of fundamental interactions are calculated;
   c) initial theses of quark screw model are given.

REFERENCES

[1]   Batsanov S.S., Zvyagina R.A. Overlap integrals and challenge of effective charges. – Novosibirsk: Nauka, 1966, – 386 p.

[2]   Fischer C.F. Average-Energy of Configuration Hartree-Fock Results for the Atoms Helium to Radon.// Atomic Data, 1972, № 4, p. 301–399.

[3]   Waber J.T., Cromer D.T. Orbital Radii of Atoms and Ions //J.Chem. Phys, 1965, v. 42, No 12, p. 4116–4123.

[4]   Clementi E., Raimondi D.L. Atomic Screening constants from S.C.F. Functions, 1.// J.Chem. Phys., 1963, v.38, No 11, p. 2686–2689.

[5]   Clementi E., Raimondi D.L. Atomik Screening Constants from S.C.F. Functions, II.//J. Chem. Phys., 1967, v.47, No 4, p. 1300 – 1307.

[6]   Gombash P. Atom statistic theory and its application. – M.: I.L., 1951, 398 p.

[7]   Clementi E. Tables of atomic functions // J.B.M. S. Res. Develop. Suppl., 1965, v. 9, No 2, p.76.

[8]   Korablev G.A. Spatial Energy Principles of Complex Structures Formation. Netherlands, Leiden, Brill Academic Publishers and VSP, 2005, 426 p. (Monograph).

[9]   Korablev G.A., Kodolov V.I., Lipanov A.M. Analog comparisons of Lagrangian and Hamiltonian functions with spatial-energy parameter.// Chemical physics and mesoscopy, URC RAS, 2004, No1, v.6, p. 5-18.

[10] Korablev G.A., Zaikov G.E. Energy of chemical bond and spatial-energy principles of hybridization of atom orbitals.//J. of Applied Polymer Science. USA, 2006, V.101, n.3, p.2101-2107.

[11] Korablev G.A., Zaikov G.E. P-Parameter as and Objective Characteristics of Electronegativity//Reactions and Properties of Monomers Polymers, Nova Science Publishers, Inc., New York, 2007, p.203-213.

[12] Korablev G.A., Zaikov G.E. Spatial-energy interactions of free radicals.// Success in gerontology, 2008, v.21, No4, p.535-563.

[13] Korablev G.A., Zaikov G.E. Formation of carbon nanostructures and spatial-energy criterion of stabilization.// Mechanics of composition materials and constructions, 2009, RAS, v.15, No 1, p.106-118.

[14] Korablev G.A., Zaikov G.E. Energy of chemical bond and spatial-energy principles of hybridization of atom orbitals.// Chemical physics, RAS, M.: 2006, v.25. No 7, p.24-28.

[15] Korablev G.A., Zaikov G.E. Calculations of activation energy of diffusion and self-diffusion// Monomers, Oligomers, Polymers, Composites and Nanocomposites Research, Nova Science Publishers, USA, 2008, pp. 441-448.

[16] Murodyan R.M. Physical and astrophysical constants and their dimensional and dimensionless combinations.// PEChAYa, M., Atomizdat, 1977, v.8, iss.1, p.175-192.

[17] Barashenkov V.S. Sections of interactions of elementary particles, M.: Nauka, 1966, 532p.

[18] Yavorsky B.M., Detlav A.A. Reference-book in physics, M., Nauka, 1968, 940p.

[19] Volkov V.V. Exchange reactions with heavy ions.// PEChAYa, M., Atomizdat, 1975, v.6, iss.4, p. 1040-1104.

[20] Bukhbinder I.L. Fundamental interactions. Sorov educational journal, №5, 1997, http://nuclphys.sinp.msu.ru/mirrors/fi.htm.

[21] Okun L.B. Weak interactions. http://www.booksite.ru/fulltext/1/001/ 008/103/ 116.htm.

In: Polymer Yearbook – 2011.
Editors: G. Zaikov, C. Sirghie et al. pp. 45-51
ISBN 978-1-61209-645-2

Chapter 4

CYCLOHEXANE OXIDATION ACTIVATED BY GLYOXAL: ELUCIDATION OF THE ACTIVATOR IMPACT

Alexander Pokutsa,[a]* Ruslan Prystanskiy,[a] Yulia Kubaj[a], Volodymyr Kopylets,[a] Andriy Zaborovskiy,[a] Jacques Muzart[b], Anatoliy Turovsky[a] and Gennadiy Zaikov[c]

[a]Department of Physical Chemistry of Combustible Fossils, Institute of Physical Organic Chemistry and Chemistry of Coal NAS of Ukraine[1], Lviv 79053. The Ukrane
[b]CNRS - Université de Reims Champagne-Ardenne, Institut de Chimie Moléculaire de Reims, UMR 6229, UFR des Sciences Exactes et Naturelles, BP 1039, 51687 Reims Cedex 2, France
N. M. Emmanuel Institute of Biochemical Physics RAS[2], 119991, Moscow, Russia

ABSTRACT

Based on the experimental data and calculated by the semiempirical method AM1 hypersurfaces of the key competition propagation reactions, it has been assumed, that prominent promoting effect of glyoxal in the cyclohexane oxidation is attributed among others to the different activity of radicals formed and the nature and structure of the interacted species as well.

Keywords: oxidation, cyclohexane, glyoxal, radicals, calculation

[1] Naukova Str., 3A, Lviv 79053. E-mail: pocutsa@org.lviv.net.
[2] 4 Kosygin Str., 119991, Moscow, Russia,E–mail: chembio@sky.chph.ras.ru,E-mail: Jacques.muzart@univ-reims.fr.

INTRODUCTION

Oxidation of cyclohexane (c-C_6H_{12}) in liquid phase [1, 2] is characterized with low (3-5%) yields in cyclohexanol and cyclohexanone. Carrying out this oxidation to conversion higher than 4-6 % under the severe conditions of the industrial process (160-180 ^{0}C, 1.3-1.5 MPa) is hampered with chain free-radical mechanism of reactions [3]. The matter is that at high level of conversion, the initial oxidation products interact with free radicals with a noticeable rate, leading to a strong selectivity decrease.

Table E_a and geometric parameters of oxidation process key propagation reactions calculated by AM1

No.	aElemental reaction	•O...H distance in TS, nm	b∠XOH, degree	Ea, кCal•mol-1
1	H2O2(-35.26) + HO• (0.63)	0.133	96.6	11.0
2	(CHO)2 (-58.75) + HO•	0.136	103.3	11.36
3	c-C6H12 (-38.62) + HO•	0.140	103.4	9.68
4	H2O2 + HC(O)C(O)OO• (-54.58)	0.124	120.5	31.56
5	(HCO)2 + HC(O)C(O)OO•	0.130	113.5	23.62
6	c-C6H12 + HC(O)C(O)OO•	0.128	114.5	25.89
7	H2O2 + HC(O)OO•(-34.83)	0.124	110.4	25.29
8	(HCO)2 + HC(O)OO•	0.128	113.4	24.71
9	c-C6H12 + HC(O)OO•	0.128	111.2	6.14
10	H2O2 + HC(O)O•(-38.60)	0.132	169.4	25.76
11	(HCO)2 + HC(O)O•	0.130	131.9	24.82
12	c-C6H12 + HC(O)O•	0.146	109.7	7.03
13	H2O2 + c-C6H11OO•(-35.37)	0.110	119.8	41.79
14	(HCO)2 + c-C6H11OO•	0.126	115.3	26.64
15	c-C6H12 + c-C6H11OO•	0.124	103.4	32.07
16	H2O2 + c-C6H11O•(-34.32)	0.122	119.7	24.98
17	(HCO)2 + c-C6H11O•	0.134	111.4	14.12
18	c-C6H12 + c-C6H11O•	0.136	112.2	16.76

[a]Superscript indexes (in brackets) display the enthalpy of formation (ΔH_f^0) of respective reactants, кCal•mol^{-1}; [b] X = C, O or H.

Studies on this problem have established that addition of small amounts of glyoxal display prominent promoting effect that allows to oxidize of c-C_6H_{12} by hydrogen peroxide in the presence of commercial VO(acac)$_2$ with high rate at 40 ^{0}C and 0.1 MPa. Conducting the oxidation under such conditions increases conversion and cyclohexanol + cyclohexanone yield (up to [x] 2), simultaneously with the enhancing of catalyst turnover number (up to [x] 10) [4]. The free radicals role in the oxidation process is displayed by the deceleration of the oxidation in the presence of a free radical inhibitor such as 2,6-*tert*-butyl-4 methylphenol [5]. It has been shown [6] that the products of glyoxal decomposition are carbon oxide, formaldehyde and hydrogen. Hence, we have admitted that, under certain circumstances, radicals from glyoxal such as HCOCO$_3^•$, HCO$_3^•$, HCO$_2^•$ are generated simultaneously to hydroxyl radicals, cyclohexylperoxy and cyclohexyloxy radicals [7], and that they can participate in the oxidation process. The present study is devoted to the elucidation and estimation of the role of these species in the key competition reactions of the chain propagation (Table).

EXPERIMENTAL

To determine the possible structure of intermediated states, elemental stages and potential energy surfaces (PES), we have applied the quantum-chemical method AM1 [8]. The correctness of calculation results by space electronic structure (SES) and enthalpy of the reactants calculated due to this semiempirical variant of MO LCAO with experimental data has been corroborated on a numerous compounds. The calculation scheme has been realized in a program package MOPAC Fugitsu (Trial version accessible from Internet). At the first stage, the solution of the task on the own value of hamiltoniane with geometry optimization by the method of BFGS [9] has been fulfilled for the cyclohexylperoxy, glyoxalperoxy, glyoxyloxy, formylperoxy, formyloxy, cyclohexyloxy and hydroxyl radicals, and molecules of cyclohexane, hydrogen peroxide and glyoxal. It should be emphasized that glyoxyloxy radicals can undergone a fast β–scission leading to CO_2 and formyl radicals. These later, in oxygen atmosphere, are instantaneously converted into formylperoxy radicals and, subsequently, formyloxy radicals. Thus, the influence of $HCOCO_2^\bullet$ and $HCO^\bullet$ radicals on the propagation reactions is negligible, and their participation has not been taken into consideration in the presented research.

RESULTS AND DISCUSSION

The most important results of quantum-chemical simulation of process elemental acts and thermodynamic properties of reactants are collected in Table. In complement to these data, it is necessary to emphasize that the maximal spin density in $HO^\bullet$, $HCOCO_3^\bullet$, $HCO_3^\bullet$, $HCO_2^\bullet$, c-$C_6H_{11}OO^\bullet$ and c-$C_6H_{11}O^\bullet$ radicals is located on p_z-AO of oxygen. The results of calculation of SES of intermediated species have been used during the determination of optimal pathways for the reaction of chain propagation. The calculation of the reaction is performed according the following protocol. The molecule which is attacked with radical was placed along the OX axis. The attacked position was determined from the $O^\bullet$...H distance and $\angle XOH$ angle (X = C, O, H) which coincides with the one in the product formed. The calculations have been conducted inside the distance interval between the reacted particles from 0.09 till 0.25 nm and the attacking angle that was changed from 60 to 180^0. The calculation step for reaction distance was equal to 0.002 nm and for attacking angle to 1^0. Unfortunately, the calculation for the whole values of angle (from 0 to 180^0) can not been performed due to steric hindrances exist between the interacted species. As the response function has been chosen the enthalpy of formation ΔH_f^0 which appeared the most sensitive to the changes in distance and attacking angle. The maximum of energy on PES is related to the transition state (TS). The last one is characterized by the distance between the interacted species and attacking $\angle XOH$ angle (Table). The noticeable energetic barrier (Table) and the presence of the clear reaction valley on the PES (Figures) undoubtedly point out that the species transformation on the course of the chain propagation reactions is passed through the TS formation.

In order to establish the thermodynamic properties of the interacted components and to determine the energetic parameter of elemental stages, the calculation of the reaction coordinate (RC) has been fulfilled by the $O^{\bullet}$...H distance with the geometric parameters optimization. RC has been designated as classical trajectory of the interacted species movement by the PES. This trajectory passes via the crest of PES and conforms to the infinity-small descent from this point into the valley of reactants or products, is begins from the pointed out maximum and the direction of RC is designated by the direction of a TS vector. Given the results of such calculations (Figure 1, Table), it can be concluded that the activation energy of interacted hydroxyl radicals with H_2O_2, $(CHO)_2$ and c-C_6H_{12} is roughly equal (Entries 1-3). Similarly, although much bigger in respect to the previous value of E_a, the activation energy of hydrogen atom breaking-off from H_2O_2, glyoxal and cyclohexane by the glyoxylperoxy (Entries 4-6) and formylperoxy (Entries 7, 8) radicals has been revealed. This difference in the E_a is anticipated and is explained with higher activity and less steric hindrance of $HO^{\bullet}$ interactions with the substrate molecules. Shifting to the formyloxy radical does not cause changing in the E_a of hydrogen atom abstraction from the inertness c-C_6H_{12} by these species (Entry 12) and is also close to the values calculated for the hydroxyl radical (Entry 3). Nevertheless, the activation energy of hydrogen atom breaking-away from $(CHO)_2$ by formylperoxy (Entries 7 and 8) and formyloxy (Entries 10 and 11) radicals appeared more than three times higher compare to c-C_6H_{12} molecule (Entries 9 and 12). According to our suggestion, the explanation of this phenomenon is concluded in noticeable steric and energetic obstacles of the last reaction caused by the structural features of $(CHO)_2$ molecule. The matter is that between the oxygen of glyoxal carbonyl group and the carbonyl group of attacking formyloxy radical a big repulsion exists. In consequence, it complicates the attack on hydrogen atoms of $(CHO)_2$ by this sort of radicals. The expressed suggestion is also corroborated by the opposite proportional dependence between E_a and value of inter-atomic distances in TS (table). For the oxy-radicals which are more active than peroxo-ones the explicable elongation of $^{\bullet}O$...H distance in TS has been revealed (Table). The properties of PES built up for all most reactive oxo-radicals, $HO^{\bullet}$, $HCO_2^{\bullet}$ and c-$C_6H_{11}O^{\bullet}$, coincide to each other and are characterized with the well-defined valley of the reaction path (Figures 1-4).

CONCLUSIONS

The higher selectivity of formyloxy and formylperoxy radicals in respect to hydrogen atoms breaking-off from the c-C_6H_{12} molecule compare to $(CHO)_2$ can be explicated by the noticeable energetic and steric hindrances in such radical attacking of dialdehyde. In consequence, the goal products yield in the $(CHO)_2$-promoted oxidation of c-C_6H_{12} by hydrogen peroxide is increased, and the process can be carried out at 40 ^{0}C with a reasonable rate.

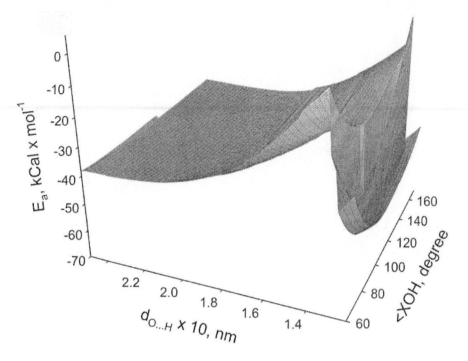

Figure 1. Potential energy surface of the reaction of chain transfer from the hydroxyl radicals on the H_2O_2 molecule.

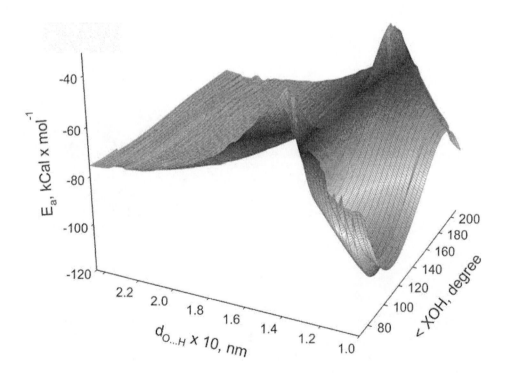

Figure 2. Potential energy surface of the reaction of chain transfer from the formyloxy radicals on the H_2O_2 molecule.

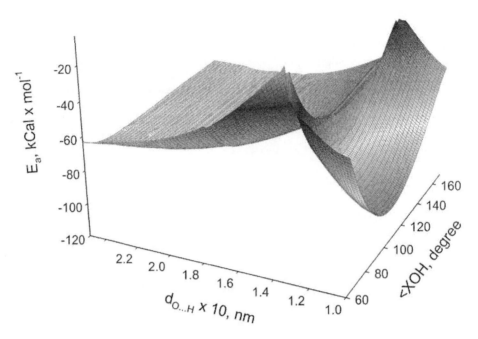

Figure 3. Potential energy surface of the reaction of chain transfer from the cyclohexyloxy radicals on the H_2O_2 molecule.

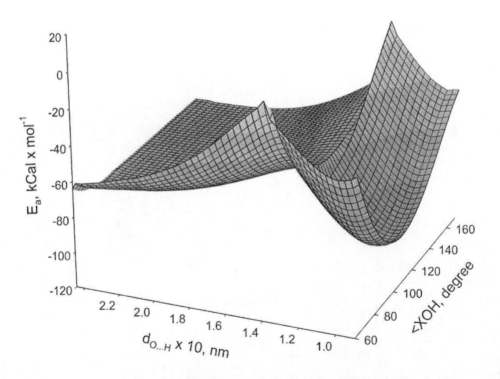

Figure 4. Potential energy surface of the reaction of chain transfer from the cyclohexyloxy radicals on the cyclohexane molecule.

The revealed function of $(CHO)_2$ coincides with the ability of aliphatic monoaldehydes to promote the oxidation of organic substrates [10]. In the case of for example propionic aldehyde, the E_a of the propionyloxy radicals interaction with $c-C_6H_{12}$ is also lower compare to the propanal [11].

This work has been supported by NATO CLG No. 982510. Authors are indebted to Prof. A. Sobkowiak for the helpful discussions.

REFERENCES

[1] Y. ISHII, S. SAKAGUCHI, T. IWAHAMA: Innovation of Hydrocarbon Oxidation with Molecular Oxygen and Related Reactions. *Adv. Synth. Catal.*, 343, 393 (2001).

[2] G. W. PARSHALL, S. D. ITTEL, Homogeneous Catalysis, Wiley, New York, 1992;

[3] N.M. EMANUEL, E. T. DENISOV: Liquid-phase oxidation of Hydrocarbons, Plenum press, New York, 1967.

[4] A POKUTSA, J. LE BRAS, J. MUZART: Glyoxal Promoted Homogeneous Catalytic Oxygenation of Cyclohexane with Hydrogen Peroxide in the Presence of V and Co Compounds. Izvestiya Akademii Nauk. Seriya Khimicheskaya, 54, 307 (2005).

[5] A POKUTSA, J. LE BRAS, J. MUZART: Benign Homogeneous Catalytic Oxidation of Cyclohexane Promoted with Glyoxal. Kinetika i Kataliz, 48, 32 (2007).

[6] F. E. BLACET, R. W. MOULTON: Photolysis of the aliphatic aldehydes. IX. Glyoxal and acetaldehyde. *J. Am. Chem. Soc.*, 63, 868 (1941).

[7] J. T. YARDLEY: Collisional quenching and photochemistry of trans-glyoxal (3Au) molecules. *J. Chem. Phys.*, 56, 6192 (1972).

[8] M. J. S. DEWAR, E. G. ZOEBISCH, E. F. Healy, Stewart J. J. P.: AM1: A New General Purpose Quantum Mechanical Molecular Model. *J. Am. Chem. Soc.*, 107, 3902 (1985).

[9] R. FLETCHER: A new approach to variable metric algorithms. *Computer J.,* 13, 317 (1970).

[10] N. KOMIYA, T. NAOTA, Y. ODA, S.-I. MURAHASHI: Aerobic oxidation of alkanes and alkenes in the presence of aldehydes catalyzed by copper salts and copper-crown ethers. *J. Mol. Catal. A: Chem.,* 117, 21 (1997).

[11] V. TIMOKHIN, M. LISOVSKA, A. POKUTSA: The Kinetics and Mechanism of Cyclohexane Oxidation by Molecular Oxygen in the Presence of Propionic Aldehyde. Kinetika i Kataliz, 41, 179 (2000).

In: Polymer Yearbook – 2011.
Editors: G. Zaikov, C. Sirghie et al. pp. 53-58

ISBN 978-1-61209-645-2
© 2011 Nova Science Publishers, Inc.

Chapter 5

TO A QUESTION OF LIQUID COMPONENTS CAPSULING

A. F. Puchkov, V. F. Kablov, V. B. Svetlichnaja,
M. P. Spiridonova and S. V. Turenko
The Volzhsky Polytechnic Institute (branch), Russia
The State Educational Institution of the Higher Vocational Training, Russia
The Volgograd State Technical University, Russia

ABSTRACT

Before use liquid components of rubber compounds can be concluded in capsules with a cover of colloidal silica. At obtaining these products (BKPITS-DBS and PRS-1) capsuling is conducted in a ball mill.

Keywords: a process of capsuling, granulated components, capsuled substances, silica, the composite antioxidant.

METHODICAL PART

The capsuling was conducted at the temperatures 40°C and 100°C in the ball mill. To the capsuling were subjected liquids with the Brukfild viscosity in 100-200 St. The viscosity was estimated with the Brukfild viscosimeter.

For study of interaction of systems an extraction was conducted in the Sokslet extractor. After completing of the extraction process the spectral analysis of residue on the filter was carried out on the infrared spectrometer, differential thermal analysis (DTA) and thermo gravimetric analysis (TGA).

RESULTS AND DISCUSSION

In technologies of polymeric materials processing preference is given to granulated components or when they are in the form of powder as in the liquid state they are inconvenient when being transported and dispensed.

Before use liquid components of rubber mixes can be concluded in capsules with a cover of colloidal silica [1, 2].

At reception of these products (BKPITS-DBS and PRS-1) capsuling is conducted in a spherical mill. BKPITS-DBS – blocked; – kaprolaktam polyisotsionat. It is used as the promotor of rubber adhesion to synthetic fibres. PRS - 1 – composite antioxidant. It is used for protection of rubbers against thermooxidizing ageing.

It is possible to present, at least, three variants of formation of capsules. The first – when silica nanaparticles settle down only on a surface of substances being capsuled (CS). It is possibly an ideal case (Figure 1). The second – when nanoparticles penetrate inside and also remain on CS surface without being aggregated (Figure 1). At last, the most real variant (Figure 1в) when nanaparticles that are inside CS there are units from nanaparticles on its surface.

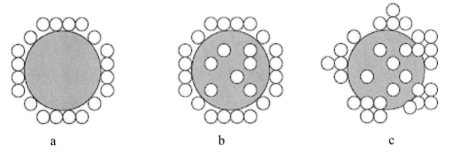

Figure 1. Formation model of capsules: a)silica nanaparticles are located on surface CS; b) silica nanaparticles take root in CS, and also remain on its surface without being aggregated; c)silica nanaparticles take root in CS, and on its surface are in a kind of units.

The elementary calculation spent on the equation (1), describing the scheme of an ideal capsule, shows, that on CS with a diameter 10 microns and density about 1 G/cm^3 it is possible to create a cover from silica particles in weight in 24.103-25.103 times as smaller as the weight of substance in a capsule provided that density of silica is 2,04 G/cm^3, and the size of its particles (accordingly and a thickness of a cover) – 25-27 nanometers. At the same time, for maintenance of real technological process and getting the product in a powder form, it is necessary to have not less than 35–50 % mass of silica with the specified size of particles. Hence, it is possible to expect that some mass of silica will enter in a capsule, and the cover will be made not of monoparticles but of their units (1), where is accordingly m_1, ρ_1, d_1 - weight, density and diameter of capsuled substances, – m_2, ρ_2, d_2 weight, density and diameter of silica nanaparticles.

$$\frac{m_1}{m_2} = \frac{1}{\pi} \frac{\rho_1 d_1^{3}}{\rho_2 d_2^{2}(d_1 + d_2)} \qquad (1)$$

Easily enough are exposed to capsuling liquids with viscosity on Brukfildu 100–200 St. In this case CS breaks into microdrops with a size 1–10 microns. More viscous liquids, at the same mechanical influences, form not equal in size and rather large particles which are difficult to consider as capsules. Such particles are capable to stick to working bodies of a spherical mill.

Questions arise – how will silica nanaparticles conduct themselves in a capsule? Whether the power condition of a capsule as a whole will change? Quite possibly that exactly silica nanaparticles will promote the course of chemical reactions.

As experiments have showed, prolonged capsuling carried out under normal conditions capable to change entropy and, to a lesser degree, the chemical nature of capsules. The course of chemical reactions becomes appreciable only at the raised temperatures.

As appears from the data resulted in table 1, durability of communication of rubber with a cord at use of the product subjected to four-hour processing in a spherical mill, is above 20 % higher of that reached at use BKPITS-DBS, capsuled during an hour.

Table 1. The influence of prolonged capsuling on modifying activity

BKPITS-DBS

Indicator	Time of capsuling, h	
	1,00	4,00
Durability of communication of rubber with a thread of a cord 23 KHTC *, the H-method, H	118	138

* In work was used soaking composition on the basis of SKD-1.

To answer a question, on what energy of blow of a sphere is spent, it is necessary to analyse thermodynamic function with reference to a capsule. Under the conditions of a constant capsule volume it is possible to express its condition using the function of free energy of Gelmgoltsa [3, 4]; F = U – T S where: U – change of internal energy, T – absolute temperature, S – entropy change. In turn, change of internal energy U = Q + A where Q – the change of the warm quantity; A – the work made over the system.

The change of the warm quantity in the process of capsuling is insignificant, allocated heat is basically dissipated, and chemical reactions in capsuled substance do not pass, which is clear from small distinction of the form of the differential thermal analysis curves (DTA) and thermo gravimetric analysis (TGA) for samples BKPITS-DBS (Figure 2), capsuled during various time testifies. The practical invariance of position of a plateau specifies in a practical invariance of structure blocked polyisotsionat.

The invariance of volume of an individual capsule indirectly testifies to absence of changes of size Δ A. Thus it is possible to assume that Δ U = 0. Then

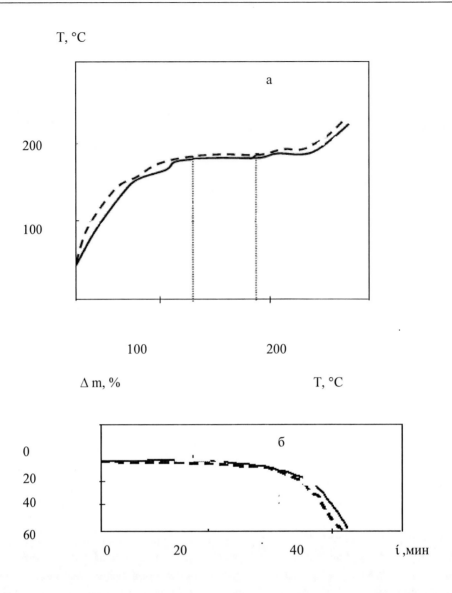

Figure2. Curves (a) DTA и (б) TGA for BKPITS-DBS, time of capsuling - 1 h (1) и 4 h (2); T1 и T2 – тemperatures of the beginning and the plateau end on curves DTA.

$\Delta F = - T \Delta S$ and, at a temperature constancy, change of free energy of system is connected only with entropy change. It is possible to assume that change of entropy of active substance in the process of capsuling, at increase in its duration, will result and in change of its entropy after the capsule will be placed in a rubber mix and is crushed by shift pressure. It, in turn, can lead to increase in probability of their interaction CS with the centres of elastomer and a cord capable for reaction.

Possibility of course of chemical reactions observed in PRS-1. CS of this product contains ε - kaprolaktam. The last one is an antioxidant of preventive action. Capsuling was spent at temperatures 40°C and 100°C. These temperatures values are defined by conditions of technological process. The extraction process (tab. 2) of substances from sample PRS-1, capsuled at 40°C comes to an end quickly enough, and, already after an hour, the quantity of extracted ε − kaprolaktam practically does not vary and makes about 97 % from initial in a composition. The extraction was spent in the Sokslet extractor.

Table 2. Kinetics of extraction of ε -kaprolaktam from the composite antioxidant PRS-1

The time of extraction min	Quantity of the extracted ε -kaprolaktam, % mass	
	capsuling at temperature 40°C	capsuling at temperature 100°C
10	18	12
20	34	27
30	65	49
60	97	88
90	97	88

It is known that to process of polymerization ε –kaprolaktam is preceded by formation amino nylon acid under the influence of a moisture [5]. Therefore, it is quite probable that 2-3 % are the quantity of ε -kaprolaktam which at presence of an adsorbated moisture silica turns in amino nylon acid, and then in polynylonamid that contacts a surface of silica particles at the expense of interaction of the carboxyl groups with the hydroxyl groups in structure Si-OH. When the extraction process is over in infra-red spectrum - a rest spectrum on the filter there is an absorption strip at 1720 cm^{-1} which, quite possibly, can belong to valency fluctuations carbonil compound-ether of the group formed at the expense of interaction hydroxide of silicon atom with carbon oxide group of polynylonamid.

Capsuling PRS-1 at 100°C leads to an intensification of the specified processes on a surface of silica particles. So, the quantity of extracted substance is about 88% (see tab. 2), and is accordingly about 12 % of products of transformation ε -kaprolaktam contacts particles of white soot. ε -Kaprolaktam connected chemically, to a lesser degree carries out the functions.

Table 3. Physicomechanical properties of vulcanized rubber mixes on the basis of rubber SKI-3

The name of an indicator	Condition of capsuling of the antioxidant	
	100^0 C	40^0 C
Conditional durability at stretching, MПa	25,0	25,0
Relative lengthening, %	650	650
Residual lengthening, %	12	12
Change of conditional durability in the course of thermooxidizing ageing, % (100 x72 h)	−23,5	−20,7

*Rubber mixes on the basis of rubber SKI-3 of a standard compounding prepared on laboratory rollers 320 160/160, the content of the antioxidant 1 mas. p., vulcanization in a press 143^0 C × 20 min.

Therefore, composite antioxidant, received through capsuling at 100°C, protects rubbers from ageing worse, than antioxidant, received at 40°C (tab. 3).

CONCLUSIONS

Thus, the blow energy of spheres of a collide mill after the capsule representing a highly-tough liquid with a cover from colloidal silica acid was generated, can promote change of entropy of system that can lead to occurrence of additional effect in increase in adhesive durability in system "rubber-cord". Thus it is impossible to exclude course of chemical reactions between silica nana particles and capsuled substance.

REFERENCES

[1]   Blocked polyisotsiants on silica / bunches Puchkov A.F., Turenko S.V., Reva S.V., Ogrel A.M. // Rubber and rubber, 2002. – № 2. – P. 23-25.
[2]   An evtektichesky antioxidant alloy besieged on silica / Puchkov A.F., Reva S.V., Spiridonov M. P//Rubber and rubber, 2002. – № 4. – P. 9-12.
[3]   Krichevsky I.R. Concept and thermodynamics bases. – M: Chemistry, 1970. – 440 p.
[4]   Muenster A. Himicheskaja thermodynamics. – M: Mir, 1971. – 295 p.
[5]   The encyclopaedia of polymers V 1. M: "The Soviet encyclopaedia", 1972. – 935 p.

In: Polymer Yearbook – 2011.
Editors: G. Zaikov, C. Sirghie et al. pp. 59-64

ISBN 978-1-61209-645-2
© 2011 Nova Science Publishers, Inc.

Chapter 6

THE STUDY PROPERTIES
OF FIRE-RETARDANT COATING

V. F. Kablov, S. N. Bondarenko and L. A. Vasilkova

Volzhsky Polytechnical Institute (branch of)
Volgograd State Technical University

ABSTRACT

The article presents the results of studies of the influence of fillers on the properties of flame retardant coating based on the phosphorusborcontaining oligomer. The influence of the content of fillers on the physical and mechanical properties of the coating has been found. The dependence of the coating properties on the amount of the filler contained in the composition has been identified.

Keywords: phosphorusborcontaining oligomer (FBO), fillers, fire protection.

PREFACE

Practice shows that the use of fire-retardant coating is one of the most economical ways to achieve the required fire resistance of building constructions. The compositions based on phosphoruscontaining oligomers are promising for the development of fire-retardant coatings. These polymer's additives are different because of a high-resistance to various external influences and under condition of relatively low concentration of phosphorus are effective flame retardants /1/. The great variety of such compounds and the possibility of changing the composition with different fillings permit to obtain a material with good performance properties.

The studying coating applies to fire-retardant intumescent paint. This is the relatively new class of materials, which attracts interest due to both their sufficiently high flame retardant efficiency and ease of use of. Under exposure to fire and heat radiation the coating forms a fire-resistant foamed heat insulating layer; its volume is several times bigger than original one /2/.

The ability of water-thinned fire-retardant coatings to produce copious carbonic foamed layer on surface of wood, which is impregnated by this coatings, under condition of ignition, illustrated in research /3/.

The information presented in researches /4-7/ considering this topic permits to interpret the properties of such compositions only from the quality point of view that makes it difficult to choose the filler and its content in the composition. So according to the things said above more careful research in this field has to be carried out. In this connection, the purpose of the work was to study the effect of fillers on properties of composites based on phosphorus or containing oligomers.

METHODICAL PART

We have used the product obtained at the Department according to the TU-40-461-806-66-07 as a phosphorusborcontaining oligomer. Fillers are wood flour (GOST - 16361 - 87), Alamo-CMC (TU 2231-034-07507908), epoxy resin (GOST 10587-84).

The ratio of components in the tested composition was as follows: 25-100 parts phosphorusborcontaining oligomer, epoxy 25-100 parts, wood flour 12,5-50 parts, Alamo-CMC 13,3-29 parts.

We investigated water absorption of samples by means of gravimetric analysis. We used a method for water absorption determination in cold water. Samples were kept in distilled water at 21 - 23 ° C during the day, then were weighed.

Intumescent factor defines the form of porous coke on the surface of sample. We investigated intumescent factor as the relative increase in the height of the coke layer after combustion against the original height of the coating.

Coke residue was determined by the relative decrease in mass of the sample, which was hold within an hour in a muffler at 600 °C.

We tested the samples for heat-resistance and we carried out mathematical analysis of experimental results by means of known methodic. We studied the behavior of materials at different temperatures by means of non-contact method using a pyrometer in the temperature range 20-500 °C.

The investigation of refractoriness of the samples was carried out by means of the method (GOST 21793-76). Refractoriness of the samples was estimated according to the time of self-extinction of samples after having been affected by free flame.

THE RESULTS DISCUSSION

The studying of water absorption permitted to make a conclusion of the following effect the quantity of epoxy resin does not influence on the water absorption. Apparently, this is due to the fact that we used an uncured epoxy resin in the compositions, and its interaction with water-soluble FBO reduced to formation of water-soluble product. The increase in the content of FBO reduces to decrease of water resistance of composition. This is due to the fact that the

FBO is well-soluble in water, and the increase in the content of FBO in the compositions leads to increase in their water solubility. Figure 1 shows the dependences of water absorption on composition.

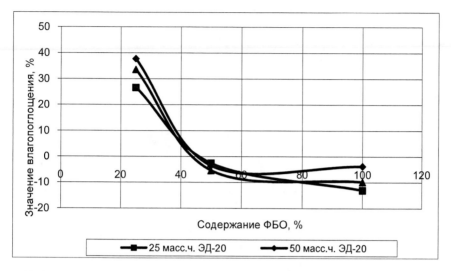

Figure 1. This is the dependence of water absorption on content of FBO.

The dependence presented in Figure 2 shows that the increase in the content of wood flour reduces to considerable increase in water absorption of samples. This is due to the fact that the wood flour is takes up water and swells well. Besides, there are compositions which don't absorb water at all. We can surmise that there is high efficiency of adsorption interaction between this field and oligomer in these samples.

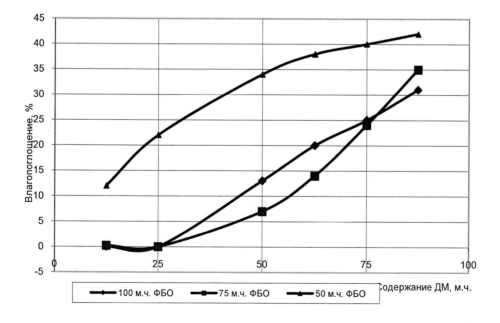

Figure 2. This is the effect of content of wood flour on water absorption.

More water containing compositions (more Alamo-CMC) are characterized by the highest rate of intumescent factor. Obviously the content of water vapor and gases released in thermally degradation influences on the significance of the intumescent factor. The compositions containing less Alamo-CMC have low rate of intumescent factor. Amount of wood flour contained in the samples affects the intumescent factor lightly. Figure 3 shows the dependences of intumescent factor on content of Alamo-CMC.

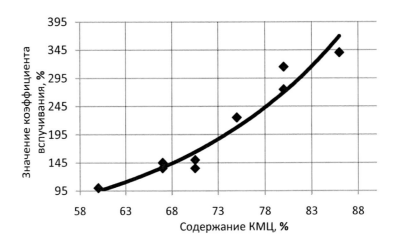

Figure 3. This is the dependences of intumescent factor on content of Alamo-CMC.

The content of phosphorus (atom) which serves as effective catalyst of coke formation considerably influences on coke production. We studied the influence of fillers on the coke formation and we revealed that the increase in the content of FBO reduces to increase in carbon residue in the compositions (Figure 4).

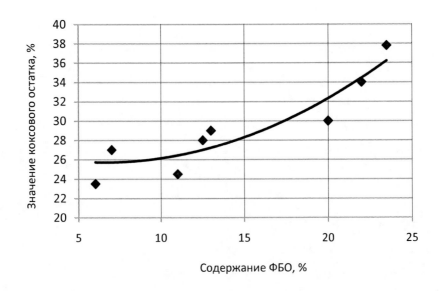

Figure 4. This is the effect of FBO content on coke formation.

The ability of fire protection coating to prevent the warming surface is an important prerequisite for its practical use. An intensive warming-up of samples during the first 2 minutes have been found out according to the data presented in Figure 5 and then its temperature doesn't change grossly. We are witnessing the achievement of a fixed sustainable temperature for each sample. This can be explained by the fact that an intensive coke formation goes on at this temperature and that prevents further heating. The samples containing the minimum amount of FBO and epoxy resin showed the best results.

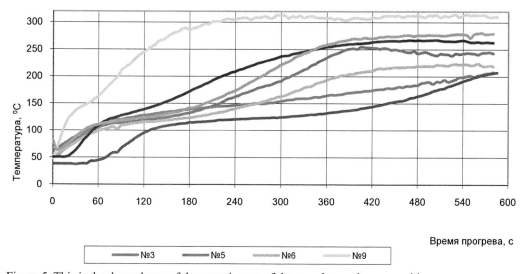

Figure 5. This is the dependence of the warming-up of the samples on the composition.

CONCLUSION

Thus, we discovered the regularities of the influence of filler content the coke formation, intumescent factor, water absorption, heat-resistance of flame retardant compositions based on phosphorusborcontaining oligomer. We found that the introduction of a large content of water-soluble fillers reduces to increase in the solubility of the composition. We showed that the introduction of the wood flour leads to significant increase in water absorption of compositions. We revealed that the top of the heating of the samples occurs at 150-200 ^{0}C regardless of the nature and content of fillers. The evaluation of heat-resistance showed that the samples containing the least amount of FBO and epoxy resin have higher heat resistance. We figured out the influence of the nature and content of fillers on the intumescent factor and we revealed that samples filled with Alamo-CMC have the increased intumescent factor.

REFERENCES

[1] Phosphorus is a fire retardant. Mode of access: http://www.sci-innov.ru/catalog/
[2] Trushkin, DV, The problems of defining combustibility of construction materials / DV Trushkin, I. Aksenov // Fire and explosion safe-sequence. - 2001. - № 4 - page. 3-8.

[3]   Kupriyanov, AV, The liquid coating materials and the technologies for their application / AV Kupriyanov / StroyPROFIl. - 2004. - № 6 - page. 25-34.

[4]   Path. 2249579 Russia. Flame retardant "OPT-HB / V. Krivtsov, I. R. Ladygina. - Declared 23/12/2003; publ. 05/06/2005.

[5]   Path. 2250204 Russia. Fire-resistant coating / EN Demin, SA Khodusov. - Declared 12/20/2002; publ. 02/08/2005.

[6]   Path. 2262523 Russia. Fire retardant coating composition / VG Kustov, G. Yu Sechin. - Declared 06/05/2003; publ. 04/12/2005.

[7]   Path. 2273616 Russia. Fire-resistant composition / MZ Kerimov, BA Suleymanov. - Declared 08/15/2003; publ. 16/10/2006.

In: Polymer Yearbook – 2011.
Editors: G. Zaikov, C. Sirghie et al. pp. 65-88

ISBN 978-1-61209-645-2
© 2011 Nova Science Publishers, Inc.

Chapter 7

PHYSICAL PRINCIPLES OF THE CONDUCTIVITY OF ELECTRICAL CONDUCTING POLYMER COMPOSITES (REVIEW)

J. N. Aneli[1], G. E.Zaikov[2], O. V. Mukbaniani[1] and C. Sirghie[3]

[1]Javakhishvili Tbilisi State University, Chemical Department,
2 Ilya Chavchavadze avue,Tbilisi-Center, Georgia
[2]N.M. Emanuel Institute of Biochemical Physics of Russian Academy of Sciences,
Moscow 119334, Russia
[3]Institutul de Cercetare-Dezvoltare-Inovare in Stiinte Tehnice si Naturale al
Universitatii "Aurel Vlaicu", Arad 310330, Romania

ABSTRACT

The role of the structural peculiarities of electrical conducting polymer composites (ECPC) has been considered. Different conception on the nature of the conductivity, the mechanisms of charge transfer in heterogeneous structures are presented in this review. Experimental results obtained by different scientists only partially are in concordance with existing theoretical models. It is suggested that missing of various physical and chemical factors influencing on the processes of electrical current formation in polymer composites is one of main reasons of mentioned divergence between theory and experimental results among which the rate of the values of inter and intra phase interactions in composites may be considered as very important factor. The peculiarities of dependence of the conductivity of systems with binary conducting fillers are considered in this work too.

Keywords: polymer composite, structure, electro-conductivity, filler content, electrical conducting polymers, interphase interactions.

---

[1] jimaneli@yahoo.com.
[2] 4 Kosygin str., Moscow 119334, Russia, Chembio@sky.chph.ras.ru
[3] Cam. 64, Corp. M, nr. 2, Elena Dragoi str., Arad 310330, Romania
Cecilias1369@yahoo.com.

INTODUCTION

Investigations of molecular and super-molecular structure effects on physical and physical-chemical properties revealed in heterogeneous polymer systems show that the formation of the structure is one of the main processes in formation of electrically conducting properties of ECPC [1-6]. In its turn, the structure significantly depends on various recipe and technological factors at production of this composites [7-9].

DEPENDENCE OF ECPC ON THE CONTENT OF FILLER

Growth of ECPC conductivity with the increase of conducting filler content is a rule without exclusions [1-4, 10]. Typical dependence of specific volume electric resistance ρ of composites, based on organic or inorganic binders, on content of conducting filler is shown in Figure1. The specific feature of this dependence is a jump-like increase of conductivity γ or, which is the same, a decrease of ρ at definite (for a particular composite) threshold filler concentration, induced by an insulator-conductor transition. This transition conforms to the so-called threshold of proceeding, or percolation. In this case γ value jump, which may reach several decimal degrees, is stipulated by formation of a continuous chain of filler particles in the polymer matrix - the infinite cluster [11,12].

Structural insulator-conductor percolation transition may be presented by a scheme (Figure 2). Resulting the increase of filler content the probability of occurrence of associates of these particles in the composition, or the so-called isolated clusters, grows (see Figure 2a). Further increase of the filler content promotes the juncture of isolated clusters into greater associates up to occurrence of an infinite cluster, i.e. a continuous electrically conducting channel in ECPC macro-system. However, in this case not all associates are included into the infinite cluster (Figure 2b). Continuous growth of the filler concentration may induce a situation, when all isolated clusters are included into an infinite cluster (Figure 2c).

In accordance with considered scheme of the infinite cluster formation, the jump-like change of ρ in Figure 1 may belong to such a concentration of the filler, when necessary conditions for occurrence of the present cluster appear. Further growth of the filler concentration leads to a monotonous decrease of ρ, followed by coming out of its values.

As it will be seen below, the transition of type insulator-conductor is sensitive to the filler content and many other factors effectively affecting the location of the filler particles.

At present the problem of the conductivity mechanism of ECPC still to be discussed. As to the opinion of some investigators [13, 14] the charge transfer is conducted by chains, consisted of filler particles having direct electric contact. On the opinion of other authors [15, 16] conductivity of ECPC is caused by thermal emission of electrons though spaces between particles. They also speak out another opinion that current exists in ECPC with air gaps or polymer films between filler particles. In this case electrons, which obtain energy below the potential barrier value may be tunneled through it, if their own wave-length is comparable with space width of insulating film [17-19].

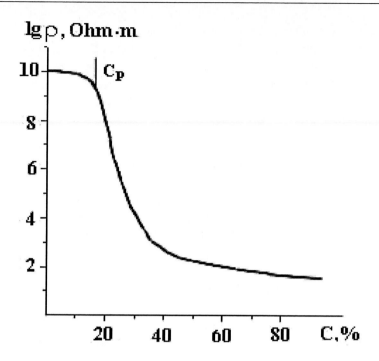

Figure 1. Typical dependence of specific volumetric electrical resistance ρ of composites on the concentration of conducting filler. Cp is the percolation threshold.

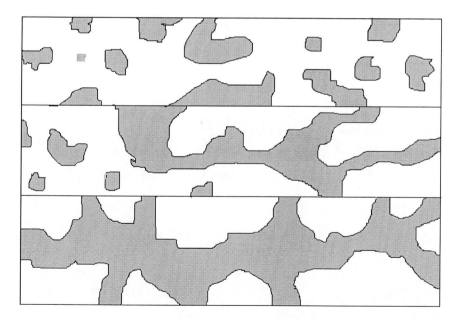

Figure 2. Scheme of infinite cluster formation from conducting particles in aniso- tropic polymer matrix.

Let us consider the most wide-spread models of the change-carrier transfer in ECPC, connected to the composition and structural features of composites.

There were the formulas suggested for calculation of electric resistance of composite, for which the formula below is the basic one for ρ calculations [20-22]. These formulas were suggested basing on the ideas of two-phase composite structures as a polymeric matrix, in which chains composed by conducting filler are dislocated according to one or another rule. In this case, it is also assumed that all conducting particles participate in formation of the electrically conducting

$$R_c = R' + R'',\qquad(1)$$

where R' and R" are electric resistance of filler particles and the sum of contact resistances between them, respectively. As total number of chains in a sample with a specific volume is

$$N = 6V_f/\pi d^2$$

where V_f and d are volumetric part and diameter of filler particles, respectively, the sum (1) could be presented as follows:

$$R = \rho_f/V_f + R_c n/N$$

Here ρ_f is the specific volumetric resistance of a filler; $R_c = \rho/2r$, where ρ is the specific volumetric resistance of the material; r is the radius of the contact point; $n = 1/d$ is the number of filler particles with diameter d.

Density packed system possesses $\rho = R_c d$ [23].

Electric conductivity of a matrix the two-phase system of a matrix (simple cubic lattice, in points of which similar sized filler particles locate) is expressed as follows [24]:

$$\gamma = \gamma_p\left[1 + \frac{V_f}{(1-V_f)/3 + V_p/(V_f - V_p)}\right],\qquad(2)$$

and electric conductivity of a statistic system (chaotic distribution of filler particles) as follows:

$$\gamma = \frac{(3V_p-1)\gamma_p + (3V_f-1)\gamma_f}{4} + \sqrt{\frac{\left[(3V_p-1)\gamma_p + (3V_f-1)\gamma_f\right]^2}{16} + \frac{\gamma_f\gamma_p}{2}},\qquad(3)$$

where γ_p and γ_f are electric conductivities of polymer and filler, respectively; V_p and V_f are their volumetric amounts, respectively.

Basing on the developed model of two-phase system conductivity the authors of suggested a formula for generalized conductivity [25]:

$$\lambda = \lambda_1\left[c^2 + v(1-c)^2 + 2vc(1-c)(vc+1-c)^{-1}\right],\qquad(4)$$

where λ is the system conductivity connected to transfer phenomenon (heat conductivity, electric conductivity, etc.); λ_1 and λ_2 are conductivities of components at $\lambda_1 < \lambda_2$; c is a parameter connected to a volumetric part of the conducting component by the expression

$$V^2 = 2c^3 - 3c^2 + 1; \ v = \lambda 1/\lambda 2$$

Some of authors think that the average distance between filler particles is a deterministic index for estimating electric conductivity of the composite [26, 27]. For example, in the case of spherical carbon particles, which form a cubic lattice in a polymer, the filler concentration will be the following [27]:

$$C = \frac{1/6 \times \pi D^3 d_f \times 100}{\left[(D+S)^3 - 1/6 \times \pi D^3\right] d_p}. \tag{5}$$

Here C is the filler concentration; S is the distance between particles; D is the diameter of particles; d_p is the polymer density; d_f is the filler density. The formula (5) makes possible calculations of the average distance S between filler particles. Similar estimation of this parameter is shown in [28].

Experimental and theoretical studies of composite conductivity were conducted in superfine gaps between graphite particles [28, 29]. In this case a significant meaning was devoted to the polymer molecule state in the gap, if filler particles were of a hypothetic form of a truncated cone. Basing on the quantum-mechanical ideas about the nature of conductivity through gaps between filler particles the following equation was deduced [30]:

$$\rho = \frac{A h^2 S}{a^2 e^2 \cdot 2 m \varphi} \times \left(1 + \beta S \varphi^2\right) \exp\left(\beta S + \frac{\beta^2 S^2 \sigma^2}{2}\right). \tag{6}$$

Here A is the parameter depending on structure of conducting particles in the system; h is the Plank constant; S is the average gap width between particles; a is the particle cross-section square; e and m are the charge and mass of electron, respectively; φ is the parameter depending on the work function of the charge yielding filler particles; $\beta = 8 \pi m \varphi / n$ is the parameter depending on the dispersion degree.

The following formula is suggested for calculating ρ [31]:

$$\lg\rho = -a\lg S + b\lg h + d_f. \tag{7}$$

Here $\lg a = n - mc$; $\lg \beta = p - qc$; $\lg d = r - tc$, where n, m, q, r, p, and t are constants; c is the mass part of the filler; S is the specific filler surface, h is hydrogen content in the filler.

According to [27]:

$$\lg\left(\rho/\rho^*\right) = \frac{\lg\left(\rho^k/\rho^*\right)}{1+\exp\left[(c-C_0/\Delta C\right]},\qquad(8)$$

where ρ^k is a specific resistance of a rubber; ρ^* is the minimum of ρ; C_0 and ΔC are the equation parameters depending on the filler type.

The authors of the work [32] suggest another formula:

$$\rho=k/c^3,\ (9)$$

where k is the parameter depending on the type of rubber; c is the filler concentration.

The paper [33] presents one more formula:

$$\rho=\exp(a/c)^p\qquad(10)$$

where a and p are constants for particular types of fillers.

In the works [34-36] the model of effective medium was used for calculation of the conductivity of ECPC possessing statistic (chaotic) distribution of conducting filler particles. This model is an analytical method of the calculation, based on the principle of the self coordination. The method is based on the supposition that calculation of electric field inside a composite element of the 'effective' medium, which conductivity is the same as the desired effective conductivity of the composite. Taking the average value of the internal field in the whole sample, it is equalized to the assigned macroscopic field. This gives the equation for determination of the effective electric conductivity [36]:

$$V_c\frac{\gamma_c-\gamma}{2\gamma+\gamma_c}+(1-V_c)\frac{\gamma_m-\gamma}{2\gamma+\gamma_m}=0,$$

where V_c is the volumetric part of the filler; γ_c and γ_m are conductivities of the filler and the matrix, respectively.

At present the percolation theory are widely used for calculations of γ for conducting composites (with both organic and inorganic binders) [11, 37]. According to this theory γ of composites consisted of non-interacting phases, may be written as follows:

$$\gamma=\begin{cases}\gamma_1\left(c_p-c\right)^{-q} & \text{at } c<c_p \\ \gamma_2\left(\gamma_1/\gamma_2\right)^s & \text{at } c=c_p \\ \gamma_2\left(c-c_p\right)^t & \text{at } c>c_p\end{cases}\qquad(11)$$

Here γ_1 and γ_2 are specific volumetric conductivities of the components; q, s and t are empiric constants (it is assumed that $\mathbf{q} = \dfrac{\mathbf{t}}{1/\mathbf{s}-1}$); c and c_p are concentration of the filler and its threshold value, respectively.

It was computed that C_p dramatically depends on the model dimension. For example, $C_p = 0.45$ for two-dimensional sample, and it is 0.15 for three-dimensional one. Another critical index t also depends on the space dimension: $t_2 = 1.3$ and $t_3 = 1.8$ [37]. However, the conditions required by the percolation theory for most of ECPC (the absence of interactions between components, first of all) are rarely fulfilled that significantly decreases possibilities of the theory application.

Recently the works [38-42] were published, in which the attempts were made to calculate the interactions between composite components. The models considered were based on the most energetically profitable states of the polymer-filler system [38]. In this case, the percolation threshold is determined, which value is different from that predicted by the percolation theory and effective medium model. The model suggested in the works [39, 40] is based on the determination of the total interphase free energy of the polymer-filler mixture. It was shown that there are other parameters, which effectively affect the formation of chain structure. They are polymer melt viscosity and diameter of filler particles. The fact is that the probability of the formation of chain structure grows with the decrease of the filler particle size. The final equation for calculation of the percolation threshold is the following:

$$\frac{1-V_p}{V_p} = \frac{3}{gd}\left[\left(\pi_f + \pi_m - 2\sqrt{\pi_f \pi_m}\right) \times \left(1 - \exp\left(-\frac{ct}{r} + K_o \exp\left(-\frac{ct}{r}\right)\right)\right)\right],$$

where g is the total interphase free energy of the mixture (polymer + filler); π_f and π_m is the surface tension of filler particles and the matrix, respectively; r is viscosity of the polymer matrix under the conditions of the composite preparation; d is the diameter of the filler particles; t is the time of mixing of two components; K_0 is the interphase free energy at the beginning of mixing (its value is determined experimentally); c is the constant of g change rate, which is also experimentally determined.

The Wessling model [41, 42] considers formation of chains as the process, based on the non-equilibrium thermodynamics. It was shown that the minimal amount of filler, which gives a possibility to obtain conducting chains, is given by the following formula:

$$C_p = 0.64(1-C)K\left[\frac{X}{\left(\sqrt{\pi_f} + \sqrt{\pi_m}\right)^2} + Y\right].$$

Here (1 - C) is the volumetric part of the amorphous fragment in the polymer matrix at room temperature; X is the constant depending on the molecular mass of the polymer; Y is the constant; K is the coefficient which calculates the presence of adsorbed polymer layers on particle surfaces.

The following formula was suggested in [43] for calculations of the γ values of ECPC:

$$\gamma = \frac{\gamma_c}{d\left(\dfrac{3}{2V_cd} - \dfrac{1}{2}\right)}.$$

Here d is the filler density in the density-packed state; γ_C is the conductivity of the density-packed cubic lattice filler particles. Values of γ calculated by this formula correlate well with experimentally obtained ones at high filling degree only (for example, for a composite of natural rubber (caoutchouc) with PME-100V carbon black).

A model of a composite structure, according to which filler particles are distributed between polymer granules (globules), allows to calculate the filler concentrations required for complete covering of globules by filler particles (V_{f1}) and formation of infinite chains in the inter globular space (V_{f2}), as well [43]:

$$V_{f1} = \frac{1}{2}P_f V_{f2} = \frac{1}{2}P_f\left(1 + \Phi\frac{r_m}{4r_f}\right)^{-1},$$

$$V_{f2} = \left(1 + \Phi\frac{r_m}{4r_f}\right)^{-1}.$$

Here I_m and I_f are the radii of polymer and filler particles, respectively; Φ is the factor depending on the type of filler particle packing and possessing the following values for different plate lattices: $\Phi = 1.110$ for hexagonal, $\Phi = 1.27$ for quadratic, and $\Phi = 1.375$ for trigonal lattice.

Nilsen et al. suggested a model of ECPC conductivity, based on polymers and metal powders [44]. In this case calculation of γ requires data about the coordination number of filler particles in the composite:

$$\gamma = \gamma_m \frac{1 + ABV_f}{1 - B\varphi V_f}.$$

Here: $B = \dfrac{\gamma/\gamma_m - 1}{\gamma/\gamma_m + A}$, $\varphi = 1 + \left(\dfrac{1 - P_f}{P_f^2}\right)$, where P_f is the coordination number of filler particles; A is the parameter depending on the particle length/diameter ratio (l/d) and the type of filler particle packing.

The works [45-49] show theoretical dependences of γ of composites with chaotically distributed fibers filler on its concentration. It was shown that γ grows with the length/-

diameter ratio of the fibers. For example, the percolation threshold for fibers with l/d = 110 equals 0.03 instead of 0.17 for spherical particles [50].

Recently, some papers appeared which mentioned that conductivity may also appear in the case, when polymer inter layers between conducting filler particles are much greater (by 3 - 5 decimal degrees) than at the current-carrier tunneling [51, 52]. It was shown that the charge transfer in ECPC is also possible at 1 nm gap between filler particles, if so-called polarons or superpolarons are formed in the polymer basing on thermodynamical profit of their formation in a polymer matrix [53, 54]. However, such systems possess non-stable electric conductivity that raises some doubts about that model of conductivity.

The model suggested in [55] determines conductivity of a composite by thickness of the polymer layer between filler particles according to the formula followed at other defined parameters of the system (work function, electron affinity to polymer, energetic structure of polymer with surface states and levels of volumetric defects in the prohibited zone, concentration and mobility of carriers, etc.):

$$a = d\left[\left(\frac{\pi(1+\varphi)}{6\varphi}\right)^{1/3} - 1\right], \tag{12}$$

where φ is the volumetric part of a carbon black in the polymer; d is the diameter of carbon black particles.

The calculation, conducted according to the equation (12), shows that d = 35 nm, if φ = 0.08, i.e. it possesses a size degree, similar to the filler particles.

The model of conductivity is shown on Figure 3. A double electric layer occurs on the border of the polymer-carbon black contact. Thickness of the charged sphere is l. At low l values (see Figure 3a) curves of the charge decrease on neighbor particles overlap, and a con-tinuous concentration of injected charge appears. This charge is able to form electric current in electric field. At high values of l (see Figure 3b) the composite possesses a sphere without injected charges. This part of the composite forms a barrier for current conduction because of its low self-conductivity (see Figure 3b).

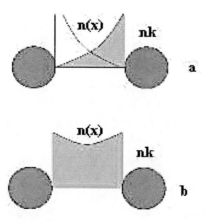

Figure 3. The model of conductivity [55].

The concentration distribution of carriers in the interlayer is the following:

$$n = n_k \left(\frac{1}{1+x} \right)^2 \tag{13}$$

$$l = \left(\frac{\varepsilon kT}{2\pi n_k e^2} \right)^{1/2} \tag{14}$$

Here n_k is the concentration of charges at the polymer-carbon black contact; e is the electron charge; x is the current coordinate; k is the Boltzman constant; T is the absolute temperature; $\varepsilon \leq$ is the dielectric permeability of the medium; 1 is the characteristic length. Such distribution of charges in the depth is usual for the case, when there are no charged traps in the prohibited zone. The criterion of ECPC conductivity is the condition a ≤ l. Substituting Eqs. (12) and (13) into Eq. (14), we obtain the following equation:

$$\varphi \geq \left\{ \frac{6}{\pi} \left[\left(\frac{\varepsilon kT}{2\pi n_k e^2 d^2} \right)^{1/2} + 1 \right]^3 - 1 \right\}^{-1} \tag{15}$$

The analysis of works on the investigations of electrically conducting properties of ECPC induces one general conclusion: despite a variety of the above considered models of electrically conducting ECPC, unfortunately no one could pretend for the versatility. Each model includes one or several approximations and suppositions, which aggravate the correctness of estimations of ECPC conducting properties. That is why the comparison of theoretically calculated data with the experimental results usually gives deviations, which reach several degrees in some cases. The coincidence is rarely reached at definite concentrations of conducting filler and specific conditions of the composite production. For example, deviation between the experimental data and those calculated by formulas (1) - (3) for ECPC, based on some thermoplastics and carbon-graphite materials, reach two decimal degrees [56]. This is apparently stipulated by an approximation of participation of all filler particles in an infinite cluster. Usually, ECPC possess γ values of separate components (of a polymer-insulator and filler-conductor, in particular), which differ by many indexes, that is why Eq. (4) displays v≈0, and than λ=λ1c2. λ is equal to 'disappearing' of the present Eq. and its transformation into a divergent function, that a high filler concentration causes significant deviations of λ values from experimental data. Similar conclusion could be made about Eq. (1.9) at high concentration of conducting filler. Great differences between computed and experimental data were also observed at the application Eq. (1.5). Apparently, it is stipulated by a limit simplification of the composite model (cubic lattice, spherical filler particles, matrix system model). Practical application of Eq. (1.6) is complicated by a significant dispersion of S and σ parameters. The necessity of experimental determination of great number of coefficients in Eq. (7) essentially decreases the degree of lasts generalization. Application of Eq. (1.1) for ρ of real composites is complicated by a wide dispersion of r values, which depend on carbon black structurization and difficulties in estimation of the interlayer thickness without preliminary selection of a mechanism for the charge transfer.

Some experimental data are satisfactorily described by Eqs. (11) [57-60]. In other cases application of this equation is correct only for rough approximations. Structural analysis and estimation of interactions between components of various electrically conducting composites shows that correctness of Eqs. (11) in relation to ECPC significantly depends on the values of interactions between components. I.e. the weaker they are, the higher is accuracy of the description of conductivity dependence on concentration, made with the help of the present equation [61-64]. It is known that the ρ values of ECPC, based on various polymers with different degree of interactions with the same electrically conducting filler at equal concentration, differ by a degree or more [65-67]. For example, ρ of chlorinated PVC and fluoroplast-based composites, filled by P357E and ATG-70 carbon blacks 35 mass part content, was found 0.25 Ohm·m and 0.036 Ohm·m, respectively [65]. In this case, it was found that comparing with PVC fluoroplast characterized by lower interaction with the filler. Values of ρ of ethylene-propylene triple copolymer and Vulcan XC-72 carbon black composite was found a decimal degree lower than that of PP-based composites with the same filler [66]. In the case of composites based on siloxane elastomer SCTV-1, ρ was found three degrees lower than for similar material with natural rubber as the polymer binder [67].

Differences in values of electric conductivity, computed according to the percolation model of conductivity and the one obtained in experiments, is frequently observed due to structural features of the filler particles. For example, experimentally measured electric conductivity of polyethylene composites, filled by acetylene carbon black, differs from the theoretical one by a decimal degree [68]. This is explained by the presence of agglomerates (associates) of particles and their statistic distribution in the matrix volume. Generally speaking, the ability of filler particles to aggregate is a significant reason of the above mentioned deviation. Difference between theoretical and experimental data on conductivity is also observed for composites, which contain a binder possessing different interaction effects with carbon black during plasticization, which is connected to free radical occurrence in this process [69]. These free radicals make their own contribution into the interaction between components. One more reason of the difficulties in the theoretical forecast of ρ value of ECPC is the existence of polar groups in macromolecules. For example, ρ of carbon black-filled composites increases in the sequence of polymers as follows: cellulose acetopropyonate < cellulose acetobutyrate < cellulose triacetate [70]. These polymers differ by hydroxyl group concentration in them, the amount of which increases in the sequence mentioned.

Comparison of different ECPC based on different thermoplastics, obtained under similar conditions, shows that the composites with crystallizing polymeric binders are characterrized by lower values of ρ, than those with amorphous binders. For example, it was shown that ρ of amorphous cis-1.4-polybutadiene, filled by "Vulkan" carbon black (in 35 mass part concentration) equals 103 Ohm·m [71]. At the same time, crystallizing trans-1,4-polybuta-diene possesses ρ = 1 Ohm·m. According to [72], ρ of the composites decreases with the growth of polyolefin crystallinity degree.

Introduction of some mineral filler (kaolin, whiting) into the composite induces growth of structural heterogeneity. This is the reason of differences in the ρ of materials with the same content of insulator (polymer + mineral filler) and conducting part [73]. Basing on the data of the structural analysis, the authors of the works [72, 73] found out that the decrease of electric conductivity in composites in the cases of either crystallizing polymers and mineral fillers is stipulated by dislocation of conducting filler particles near the surfaces of crystallites or kaolin and other mineral fillers consequently to more dense packing of current-conducting channels in amorphous (lower dense phase) of the polymer. However, some authors with no reason ascribe this experimental result to high conductivity of crystalline forms in polymer [71, 74].

Taking into account interactions between phases and in phases of ECPC we obtain satisfactory results by using formulas (16) for ECPC with completely amorphous binder [56]:

$$\rho = \rho_0 \exp\left(\frac{C_p - C}{C_p}\right)^a \quad \text{at } C > C_p$$

$$\rho = \rho_0 \qquad\qquad \text{at } C \leq C_p \,, \tag{16}$$

where ρ_0 is the specific volumetric electrical conductivity of pure polymer, equal to ρ of the composites containing a conducting filler in concentrations below the percolation threshold mass part ($C \angle C_p$); a is the constant proportional to the expression: a ~ $\dfrac{e_2 \times e_3}{e_1^2}$, and depending on the energies of interactions of polymer-polymer (e1), polymer-filler (e2) and filler-filler (e3) types.

Analysis of the ideas suggested in [10, 56] on the influence of the ratio of interaction energies between components of the composites induce a conclusion that the situation, when e2 and e3 are close by values and e1 reaches its minimum. In this case, ρ also obtains its minimum, whereas growth of any of e2 and e3 induces the increase of ρ [10]. If e2 > e3, the probability of stable bond formation between filler particles decreases, i.e. the system looses its conductivity. But if e2 < e3, the probability of agglomerate formation from conducting particles grows that leads to a decrease of branching of conducting channels (pathways). In both cases we obtain growth of ρ.

It is evident that preliminary estimation of energetic parameters e1, e2 and e3 is very difficult (estimation of the affinity between components by adhesive parameters), but conduction of several experiments can give the parameter a for the components of the present composite that significantly simplifies calculation of ρ for different concentrations of conducting filler in the same ECPC by equation (16). Application of this equation to polymeric composite, the polymer phase of which contains crystalline spheres, is also possible in the case, when the mass part of the binder sums up only the amorphous part of the polymer, in which filler particles are localized.

To clear up the correctness of Eq. (16) application in ρ calculation and comparing it with the experimental data, the tests of electrically conducting rubbers, based on organosilicon elastomers of type SCTV (polydimethylmethylvinilsiloxan) and three types of carbon black P803, P357E and ATG-70, were conducted [56]. All samples were obtained by the additive vulcanization technique with ADE-3 (diethyl-aminomethyl-triethoxisilan) as hardener (curing agent). The main difference between these types of carbon blacks is in values of specific geometrical surface S and ρ (the ρ values for these carbon blacks were found 14·10-4, 25·10-4 and 1.6·10-4 Ohm·m, respectively; S values were 106, 56 and 46 m2/g, respectively). Materials possessing different ρ values were obtained by introduction of different amounts of the mentioned carbon blacks into composites. Figure 4 shows that the character of the ρ dependence on the filler concentration significantly depends on the filler type. For example, to obtain rubbers containing P357E and P803 carbon blacks and possessing equal ρ values, significantly greater amounts of P803 should be introduced comparing with P357E.

The result obtained correlates well with the data from [75], which show that a sufficient effect on ECPC conductivity is induced by the carbon black dispersion and the ratio of carbon black particle square to its mass (S/m). The value of γ of ECPC containing carbon blacks with different S/m values increases proportionally to this ratio with the concentration.

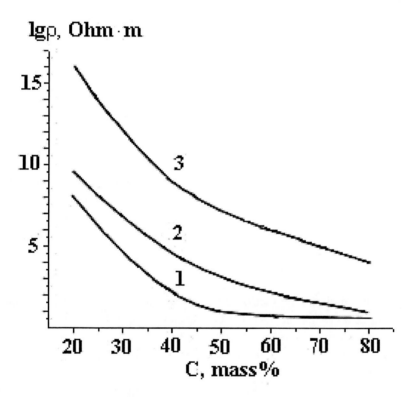

Figure 4. The dependence of ρ values of SCTV-based composites on the filler concentration. The fillers are P357E (1), ATG-70 (2), P803 (3).

Table 1 shows experimental data and the results of the ρ value calculations by Eq. (16) for SCTV-based composites with various filler contents, and the filler concentration C_p, corresponded to the insulator-conductor transition for the same materials. To estimate generality degree of the formula (16) ρ and C_p, were also calculated for non-organosilicon conducting rubbers and compared with experimental data on those materials, obtained by different authors.

**Table 1. Experimental and calculated data on ρ and Cp
for electrically conducting rubbers**

Composite	ρ_{exp}, Ohm·m	ρ_{calc}, Ohm·m	C_p (exp.)	C_p (calc.)	Ref.
SCTV + P357E (40)*	0.058	0.045	12	10	56
SCTV +ATG-70 (50)	0.04	0.03	16	13	56
SCTV +P803 (60)	0.19	0.16	40	48	56
BSC + Vulcan-3 (50)	25.3	22.4	30	35	69
NC + ATG-70 (50)	18.7	19.6	28	33	61
SCN + PM-100 (60)	11.6	13.8	25	30	3

- numbers in brackets mark mass parts of the filler per 100 mass parts of elastomer.
C_p is measured in the same units.

The data shown in Table 1 display that deviation between experimental and calculated data not exceed 20%. In this case we can state that Eq. (16) may be used for calculations of concentrational dependences of ρ in ECPC with amorphous polymeric matrix.

STRUCTURE MODELS OF ECPC

Conductivity of polymers, filled by electrically conducting fillers, depends, first of all, on the current-conducting channel density in a polymer matrix that, in its turn, seriously depends on capability of filler particles for forming an infinite cluster. It was mentioned above that formation of a current-conducting system in polymer sufficiently depends on the ratio of interaction energies between the composite components. If we take into account that highly structured carbon blacks P357E and ATG-70 possess comparatively high energies of interactions between their own particles with polymer, and that intermolecular interaction in organosilicon elastomers is weaker than in other polymers, it becomes clear why composites based on highly structural carbon blacks and SCTV possess the conductivity higher than of the composite, based on SCTV and lower structural carbon black P803.

Physics and chemistry of the surface of filler particles are the decisive measures in the filler-polymer and filler-filler interactions that, in its turn, play the leading role in formation of the structure and electrically conducting properties of ECPC [2, 10].

The structure of carbon black and graphite seriously affects the electric conductivity of composites. In some cases, the increase of the structure degree becomes more effective, than the increase of specific surface square. For example, rubbers filled by higher structured carbon black (PM-90) possess higher conductivity, than those filled by lower structured but higher dispersed carbon black (PM-100) [76]. Similar result was obtained for the comparison of the conductivity of conducting rubbers, filled by highly structured acetylene carbon black ATG-70 and lower structured PM-100 [77]. However, the situation often occurs, when the effectiveness of carbon blacks is compared with other inter compensating properties (structure degree, dispersion, porosity, roughness, etc.), which complicate estimation of one or another factor. It is known that dispersion [78] and porosity significantly affect conductivity of filled rubbers and plastics. The analysis of effects of structural indexes of carbon blacks on electric conductivity of composites is presented in [79].

Chemical composition of carbon black particle surfaces is very important for the analysis of the carbon black type effects on the conductivity of ECPC. Substances, adsorbed or chemically bonded to surfaces of carbon blacks, may prevent formation of contacts between particles or promote formation of bonds between polymer and fillers. Chemical properties of the surface are defined by the existence of functional groups, consisted of oxygen, hydrogen, sulfur. Amounts of oxygen and hydrogen in carbon blacks reach 5% of carbon mass. Oxygen exists in the basic composition of carboxylic, phenolic, quinoid and lactic groups. Many data support the idea about free-radical origin of carbon blacks [80, 81]. Destruction of carbon black structure is an additional source of free radicals, which significantly affect further filler interactions with the polymer [82]. The effect of functional carbon black groups on affinity to the polymer depends on the polymer nature. For example, its adhesion to butylcaoutchouc increases at carbon black oxidation, and adhesion to BSC and polybutadiene decreases [3].

Preliminary thermal treatment of a carbon black in inert atmosphere at high temperatures (over 1000 K) effectively affects conductivity of composites. Experiments showed that in most cases conductivity of ECPC containing heat-treated carbon blacks increases (in some cases by 6 decimal degrees) [83].

The chemical groups of carbon black surface significantly influence on the polymer-filler interactions, because it may cause an activation of different types of interactions. High energy of the polymer-filler interaction may promote the ECPC structure degradation. Oxidation of carbon black particle surfaces always increases ρ, and elimination of volatile substances and chemical groups at thermal treatment without oxygen induces ρ decrease in ECPC [84].

Influence of conducting filler type on the percolation threshold is well seen in the investtigation of electrically conducting properties of polyester epoxy-based composites, dissolved in styrene with carbon-graphite fillers [85]. Hydroperoxide of isopropyl benzene oxide (hyperysis) is the hardener of that composite, and cobalt naphthenate is the accelerator of the process. Mixtures were prepares according to two techniques: by mixing ingredients in a vessel with a mixer (high-ohmic samples) and cold pressing of previously rolled masses in press-forms under 15 Mpa pressure (low-ohmic samples). The choice of preparation technique depends on viscosity of mixtures, which, in its turn, depends on the filler concentration. Low concentration of the filler and, consequently, low viscosity of the mixture, induces a possibility of mixture preparation in an usual mixer with mechanical mixing machine. Increased filler concentration and viscosity require significant mechanical forces and application of rollers.

The compositions, produced in accord with the mentioned technique, differ by an intensive increase of conductivity at comparatively low filler concentrations. The data from [3] say that similar transition in different composites occurs at relatively high filler concentrations. For example, this fact is explained in the work [19] by an irregular distribution of the filler in the polymer matrix. Microstructure of the composite represents electrically conducting spatial network, consisted of the filler particles, and disposed between dielectric blocks (domains). These blocks may be formed by macromolecules with definite order in the distribution of ones or crystal areas (Figure 5). Polymer blocks (domains) may be formed as a result of macromolecular aggregation via interactions under the effect of Van-der-Waals forces and electrostatic forces of polar groups (some authors named such blocks "the minor elements of supramolecular structures" (NENS) [86]. Particles of electrically conducting filler form a conducting structure, concentrated in the inter-block space. This structure appears denser, than it would be in the case of the block structure absence. Thus, morphological features of the considered composition are the deterministic factors of the conducting channel formation with relatively low threshold concentration of the filler [87]. In this case, the effect of the filler structure degree correlates with the experimental data in conductivity dependence on the carbon black type [18].

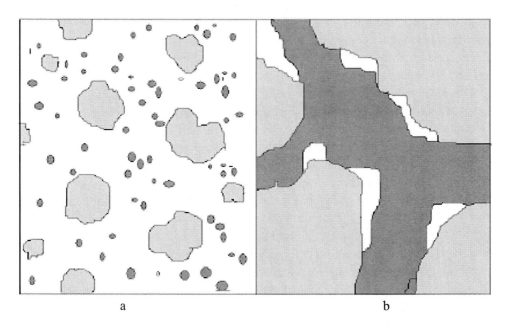

a   b

Figure 5. The scheme of current-conducting system (in accordance with an electron microscopic picture) formation on the base of polymers and electric conducting particles. a- initial state before formation of infinite clusters in ECPC (dark spots- conducting particles, light spots – polymer domains, white area -amorphous polymer); b – ECPC after formation of the infinite cluster (dark area) among polymer domains (light area) and free volume (white area).

Application of two different types of electrically conducting fillers in a single composite induces an extreme character (with a minimum) of ρ in accord with the ratio of the fillers. The works [85, 88] show the curves of ρ dependence on concentrations of two conducting fillers: graphite and carbon black, at various concentrations, and the following equation for calculation was suggested:

$$\rho = \frac{ac_2^2}{\left(c_1+c_2\right)^m} - \frac{bc_2}{\left(c_1+c_2\right)^n} - \frac{K}{c_1^3}. \tag{17}$$

Here c_1 and c_2 are concentrations of ATG-70 and graphite, respectively; a, b, K, m, n are coefficients depending on the type of elastomer. At $c_2 = 0$ Eq. (17) transforms into (9). Although the authors of [87, 89] succeeded in the application of Eq. (17) for ρ calculation for various combinations of binary filler components at different total filling of SCI-3-based rubber, that equation is a one-side playing mean, because it displays no invariance to binary filler components. Moreover, it was mentioned above that the consideration of allied Eq. (1.9) displays incorrectness of Eq. (17) at transition from specific volumetric resistance to specific volumetric electric conductivity of the material.

To clear up the functional dependence of ρ of ECPC on concentration of the binary filler, polyester varnish-based composites with carbon-graphite filler were produced [84] (C-1 graphite and P357E, ATG-70, and P803 carbon blacks). The composites with P803 and graphite (total concentration was 40 mass parts) displayed the change of ρ displayed by a curve with a minimum, corresponded to P803 carbon black concentration of 25 mass parts and graphite of 15 mass parts (Figure 6). It is known [10] that carbon black is capable for creating a secondary structure owing to the existence of an active surface as associates of particles or clusters, that leads to the formation of a three-dimension conducting system. Possessing relatively high conductivity, graphite displays no such a capability. That is why composites containing carbon black as filler are characterized by much higher conductivity, than the composites based on the same polymer, filled by the same graphite amount. Figure6 shows curves reflecting one of the dependences of the so-called synergic effect. This effect concludes in inhomogeneous distribution display at the increase or weakening of another reactions and propertiesof material at introducing two or more active components into it.

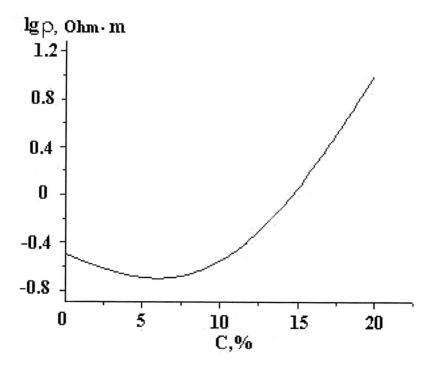

Figure 6. The dependence of ρ of polyester epoxy-based composites on the ratio of binary filler (graphite + P803) components at the sum content of fillers 20%. On the x-axis – the content of graphite in fillers blend.

Synergism of binary fillers is connected with the features of the composite morphology. In particular, this phenomenon is explained by the type of inter disposition of two type filler particles in the polymer matrix. For example, microstructure of a composite, which contains carbon black and graphite, may be schematically presented as a conglomerate of particles of the fillers, 'injected' into the polymer matrix (Figure 7). Carbon black particles possessing lower electric conductivity form a secondary structure, looking like bridges between conducting particles of graphite, including them into the general conductive system. If it is presented as an electrical scheme of parallel-consequent connected resistant elements, it becomes possible to explain the reason of a significant improvement of electrically conducting properties of the composite.

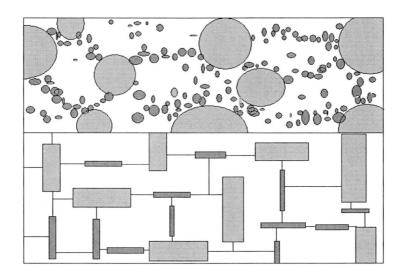

Figure 7. Two-dimension model of the composition based on polymers with a binary filler (graphite + carbon black) (top) and equivalent direct current scheme (bottom). Big circles - graphite particles; small circles - carbon black particles;big rectangles - resistance of graphite particles; small rectangles - resistance of carbon black particles.

The experimental data on electric conductivity of ECPC with binary electrically conducting filler at different values of total filler concentration and simultaneous application of mathematical planning of the experiment [90] allow setting regularity for the ρ-c dependence, described by the following formula:

$$\rho = -A(\rho_1 c_1 \ln c_2 + \rho_2 c_2 \ln c_1) \tag{18}$$

Here ρ_1 and ρ_2 are specific volumetric resistances of pure fillers (carbon black and graphite), respectively; c_1 and c_2 are concentrations of these fillers in mass parts; A is the constant depending on the material type. Calculations by Eq.(18) should be easier conducted for one concentration, i.e. expressing the second filler concentration via the first one, taking into account that $c_1 + c_2 = 1$:

$$\rho = -A[\rho_1 c_1 \ln(1-c_1) + \rho_2 (1-c_1)\ln c_1] \tag{19}$$

Experiments showed that the data on the determination of the ρ dependence on composition of the binary filler (carbon black 1 + carbon black 2, graphite + carbon black) satisfactorily correlate with those calculated by Eq. (19).

Influence of the composition on electrically conducting properties of ECPC was shown on the example of the systems, consisted of two types of organosilicon elastomers SCTVF-803 and SCTVF-2103 and carbon black fillers P803, P324 and ATG-70 [91]. Some of the composites contained A-300 aerosil. Concentrations of the fillers were varied from 20 to 80 mass parts per 100 mass parts of elastomer. Dicumyl peroxide in 3 mass part concentrations was used as a vulcanizing agent. Rubber mixtures were prepared on laboratory rolls. Vulcanization was performed by the well-known technique of peroxide vulcanization [92]. Electrodes were introduced into the rubber mass before the vulcanization began.

It should be mentioned that in most cases investigators measure electric resistance of materials by the four electrode technique [3].

Table 3 shows characteristics, obtained in tests of electroconducting and physical-mechanical properties of vulcanizates [56].

**Table 3. Physical and mechanical indexes of electroconducting rubbers,
based on SCTVF-803 and SCTVF-2103 elastomers**

Group	N	Composite	ρ, Ohm·m	σ, MPa	ε, %	θ, %
I	1	SCTVF-803 + P324 (30)*	50	3.3	200	0
	2	SCTVF-803 + P324 (50)	11	6.2	260	8
	3	SCTVF-803+ P324 (60)	0.42	3.8	200	16
	4	SCTVF-803+ P324 (80)	0.37	2.2	120	26
II	5	SCTVF-803 + P803 (40)	10^9	2.0	160	0
	6	SCTVF-803 + P803 (50)	21	3.0	140	0
	7	SCTVF-803 + P803 (60)	2.4	4.1	140	0
	8	SCTVF-803 + P803 (80)	0.7	4.6	100	0
III	9	SCTVF-803 + P324 (20) + A300(20)	300	7.7	350	4
	10	SCTVF-803 + P324 (30)	48	6.3	220	3
	11	SCTVF-803 + P324 (40)	17	6.1	220	8
IV	12	SCTVF-803 + ATG70(20) +A300 (20)	37	6.4	270	4
	13	SCTVF-803 +ATG70(25)	1.2	6.5	250	4
	14	SCTVF-803 +ATG70(30)	0.25	6.5	230	8
V	15	SCTVF-2103 + P803(60)	9.9	4.0	100	0
	16	SCTVF-2103 + P324(60)	7.3	4.9	265	15
	17	SCTVF-2103+ATG70(25) +A300(20)	0.7	6.6	280	4
	18	SCTVF-2103+ATG70(25) +A300(40)	1.8	8.4	175	4

*Numbers in brackets correspond to filler concentrations in mass parts per 100 mass parts of the elastomer.

According to the data shown in Table 2, SCTVF-803-based composites possess higher conductivity, than SCTVF-2103-based rubbers at equal concentration of the filler. For example, ρ of SCTVF-803-based rubber, which contains 60 mass parts of P324 carbon black, is one decimal degree lower, than that of SCTVF-2103-based one containing the same filler in the same concentration. This may be explained on the basis of two phenomena: 1) Filler dispersion during rubber mixture rolling and 2) Distribution of filler particles in elastomer matrix.

Growth of the filler concentration, induced by dispersion increase, promotes simultaneous growth of the number of conducting channels and, consequently, the decrease of ρ, only in the case if electrically conducting particles form branched spatial network in the matrix. This becomes possible at a definite ratio of intensities of two types of interactions: the filler-filler and the elastomer-filler interaction. Predomination of the first type of interaction ambiguously induces the increase of the rubber conductivity, because in this case the formation of associates (lumps) is intensified. These lumps induce increase of in homogeneity in the filler particle distribution that might cause the ρ growth and decrease of physical and mechanical parameters starting from a definite (for the particular composite) filler concentration. That is why the existence of the elastomer-filler interaction is also required for the formation of highly developed conducting system in the rubber. This interaction prevents the process of the lump formation. Consequently, one may suppose that high conductivity is obtained by composites at a definite ratio of the mentioned interactions.

Taking into account the supposition and technical indexes, shown in Table 2, the ratio of the interactions mentioned in the first of the compared composites (among No. 3 and No. 16 composites) should be optimal comparing with the second one (ρ of the first composite is lower than that of the second one). The spin probe technique was used for obtaining results on the homogeneity degree. It was found that the homogeneity degree of the filler distribution in the matrix of No. 3 rubber is lower, than that in the composite No. 16. This correlates well with the known character of the filler particle distribution in composites with high compatibility of the components [93].

The effect of the filler type on the properties of composites are well seen on the example of two groups of rubbers, based on SCTVF-803 elastomer with two types of carbon black (P324 and P803). Carbon black P324 possesses higher conductivity than P803. That is why these composites possess different ρ values. However, it should be taken into account that the difference in properties of the composites of the groups I and II is stipulated by the properties of carbon blacks separately, and by their behavior in the polymer matrix. This affects, in particular, the physical and mechanical indexes of the composites. For example, if the maximum of resistance of the group I rubbers is displayed at 50 mass parts concentration of the carbon black P324, the rubbers of the group II possess the maximum (according to the tendency of resistance growth) at higher filler concentrations. Moreover, difference in the properties of the groups of composites compared is also expressed by the value of residual elongation: all composites of the group II are characterized by the absence of it.

The reason of the mentioned differences in the properties of those groups of rubbers should be searched in the character of interactions between the composite components. On the one hand, stronger polymer-filler and filler-filler interactions in the rubbers of the group I, comparing with the group II, induce higher conductivity, and on the other hand, they promote formation of a composite with the maximum resistance at relatively low filler concentrations. Zero values of the residual elongation of the group II rubbers evidently point out fast relaxation processes in the macromolecular system, which proceed in the composites after the sample rupture, caused by a weak polymer-filler interaction.

Dielectric filler aerosil is known as a good intensifier of rubber mixture [7]. That is why in the obtained three-component systems aerosil A-300 acts as an intensifier of organosilicon rubbers (groups III and IV). However in the case of the present filler, optimal concentrations do also exist, which give high physical and mechanical properties to rubbers. For example, the sample with lower concentration of the binary filler aerosil + carbon black is characterized by higher resistivity (sample 9), than the sample with higher carbon black concentration. The improvement of electrically conducting and resistive properties of composites is observed at ATG-70 carbon black application, combined with aerosil (samples from the groups IV and V) at optimal ratio of the fillers. Thus, variation of the filler concentration may improve some properties at simultaneous decrease of others. For example, the increase of aerosil concentration induces the decrease of electric conductivity of the composites with binary fillers, but resistance simultaneously increases (samples 17 and 18). Aerosil effect is evident and requires no additional explanations. In its turn, decrease of the conductivity of rubbers at aerosil concentration growth may depend on two factors: the decrease of the total part of conducting filler in the composite and destruction of the current-conducting system by aerosil particles.

CONCLUSIONS

The experimental data confirm that most important factors effectively influencing on the conductivity of ECPC are following: the concentration, average size and type of filler particles and value of three type of interactions: macromolecule-macromolecule, macromolecule-filler and filler-filler.

Searching for the ρ dependence on the filler concentration in ECPC should, probably, induce a logic conclusion that a composition with highest conductivity may be obtained at the maximal filling degree. There is no doubt in this, but it is also known that technologists are induced by deterioration of physical and mechanical properties of the composites at high filling degree to introduce some limits for selection the optimum concentration of conducting fillers.

REFERENCES

[1]   R.N. NORMAN: Conductive Rubber and Plastics, Amsterdam, Elsevier, (1970).

[2]   A. DONNET, A. VOET: Carbon Black, Marcell Decker, New-York-Basel, (1976)

[3]   V.E. GUL', L.Z. SHENFIL: Electrically Conductive Polymeric Composites, Moscow, Khimia, 240 (1984) (in Russian).

[4]   4.CARBON BLACK POLYMER COMPOSITES: The Physics of Electrically Conducting Composites, Marcell Decker, New-York, (1982).

[5]   J.N.ANELI, L.M.KHANANASHVILI, G.E.ZAIKOV, Structuring and Conductivity of Polymer Composites, Novo-Sci. Publ., New-York, 1998.

[6]   V.S. Krikorov, L.A. Kalmakova: Electrically Conductive Polymeric Materials, Moscow, Khimia, 176 (1984) (in Russian).

[7]   F.F. Koshelev, A.E. Kornev, N.S. Klimov: General Rubber Technology, Moscow, Khimia, 560 (1968) (in Russian).

[8]   A.A. Berlin, S.A. Volfson, V.G. Oshmyan, N.S. Enikolopov: The Principles of Creation of Composite Polymeric Materials, Moscow, Khimia, 240 (1990) (in Russian).

[9]   Y.S. Lipatov, Physical Chemistry of Filled Polymers, Moscow, Khimia, 304 (1977) (in Russian).

[10]   K.A. Pechkovskaya: Carbon Black as Rubber Reinforcement, Moscow, Khimia, 304 (1968) (in Russian).

[11]   B.I. Shklovski, A.L. Efros: Electronic Properties of Alloyed Semiconductors, Moscow, Nauka, 416 (1979) (in Russian).

[12]   G. Clero, T., Giroult, I., Russank: Acad. Sci. Comptes Rendus, Ser.B, 281 (13), 227 (1975).

[13]   Z.PETROVICH,B.MARTINOVICH,V.DIVIYAKOVICH   AND   J.BUDINSKI-SIMENDICH, J.Appl.Pol.Sci.49, 1659 (1993)

[14]   L.BENGIUIGUI, J.JAKUBOVICH, M.NARKIS, J.Pol.Sci., Part B: Polymer Physics, 25,127(1987).

[15]   F.F. Koshelev, A.E. Kornev, E.M., Spiridonova: Electrically Conductive Polymeric Materials, Moscow, CBTI, 61 (1961) (in Russian).

[16]   L.K. VAN BEEK , B.I. VAN PUL: Carbon, 2, 121 (1964).

[17]   G.BEAUCAGE, S.RANE, D.W.SHAFFER, G.LONG, D.FISHER, J.Pol.Sci.,Part B,37,1105 (1999).

[18]   B. Lee: Polymer Eng.Sci., 32(1), 36 (1992).

[19]   W.B. Bridge, M.J. Folkes, B.R. Wood: J.Phys.D., 23(7), 890(1990).

[20]   B.I. Sazhin: Electrical Conductivity of Polymers, Moscow, Khimia, 376 (1970) (in Russian).

[21]   E.M. Dannenberg: SPE J., 21(7), 36 (1965) .

[22]   YIHU SONG, QIANG ZHENG, J.Appl.Pol.Sci.,105,710,(2007).

[23]   V.N. Anikeev, V.S. Zhuravlev: Colloidni Zh., 46(6), 1157 (1979) (in Russian).

[24]   V.I. Odelevski: Zhurn. Thech. Phys., , V.21, N6, p.667-685 (1951) (in Russian).

[25]   G.N. Dulnev, V.V. Novikov: Transfer Processes in Inhomogeneous media, Moscow, Energoatomizdat, 248 (1991) (in Russian).

[26]   M.N. Pooley, B.B. Boonstra: Rubber Chem.Technol., 30(1), 170 (1957).

[27]   E.M. Abdel-Bary, M. Amin, R.H.Hassan: J.Polym.Sci., Polym. Chem. Ed., 17 (7), 2163 (1979).

[28]   N.S. Enikolopyan, S.G. Gruzova, N.M. Galashina, E.N. Sklyarov, L.N. Grigorov: Dokl. AN USSR, 274(6), 1404 (1984) (in Russian).

[29]   L.N. Grigorov, N.M.Galashina, N.S. Enikolopyan: Dokl. AN USSR, 274 (4), 840 (1984) (in Russian).

[30]   K. Ohe, G. Natio: Jap.J.Appl.Phys., 10(1), 94 (1971).

[31]   M.L. Studebacker: India Rubber World, 129(4), 485 (1954).

[32]   V.V. Pushkova, Yu.F. Kabanov, M.I. Kulakova, V.I. Gudimenko, A.M. Lukyanov, I.A. Ragozina, E.V. Minervin: Kozhanno-obuvnaya promishlennost, 5, 39 (1971).

[33]   B.B.BOONSTRA, Rubber Chem.Technol.,50(1),194 (1977).

[34]   A.K.ZAINUTDINOV, A.A.KASIMOV and M.A.MAGRUPOV, Pisma v zhurnal eqsperimentalnoi i tekhnicheskoi fiziki, 18(2),29(1992).

[35]   R. Landauer: J.Appl.Phys., 23, 779 (1952).

[36] B.N. BUDTOV, Y.I.VASILENOK, V.V.VOITOV, A.A.TRUSOV, Fizika Tverdogo Tela, 31(8),262(1989).

[37] B.I. Shklovski, A.L. Efros: Uspekhi Fizicheskikh Nauk, 117(3), 401 (1975) (in Russian).

[38] T. Lux: J.Mater.Sci., 28, 285 (1993).

[39] K. Mijasaka, K. Watanabe, C. Jiojima, S. Asai, M. Sumita, K. Ishikava: J.Mater.Sci, 17, 1610 (1982).

[40] M. Sumita, K. Sakata, S. Asai: Polymer Bull., 25, 265 (1991).

[41] B. Wessling: Macromol.Chem., 185, 1265 (1984).

[42] B. Wessling: Synt.Metals, 28, 849 (1989).

[43] A. Malliamis, D.T. Turner: J.Appl.Phys., 42(2), 614 (1971).

[44] L.E. Nielsen: Ing.Chem.Fund., 13, 17 (1974).

[45] G.E.Pike, C.H. Seager: Phys.Rev.,B, 110(4), 1421 (1974).

[46] J. Yamaki, O. Maeda, Y. Katayama: Rev.of Electrical Commun. Lab., 26(3/4), 610 (1978).

[47] J. Yamaki, O. Maeda, Y. Katayama: Kobunsi Ronbunsi, , 32(1), 42 (1975).

[48] S.M. Musamoto, M.K. Abdelazeez, M.S. Ahun: Mater. Sci. Eng., 10(B), 29 (1991).

[49] E. Cherleux, E. Gugon, W. Rivier, M.S. Ahun: Solid State Comm.,50(11), 999 (1984).

[50] F. Carmona, R. Cauct, P. Delhas: J. Appl. Phys., 61(7), 2550 (1987).

[51] B. Bridge, M.I. Folkes, H.D.S. Jahahani: J. Mater. Sci., 23, 1955 (1988).

[52] B. Bridge, M.I. Folkes, H.D.S. Jahahani: J. Mater. Sci., 25, 3061 (1990).

[53] L.N. Grigorov: Visokomol. Soed., 27A(5), 1098 (1985) (in Russian).

[54] S.G. Smirnova, L.N. Grigorov, N.M. Galashina, N.S. Enikolopyan: Dokl. AN USSR, 285(1), 176 (1985) (in Russian).

[55] A.P. Losoto, Yu.M. Budnitski, M.S. Akutin, A.T. Ponomarenko, A.A. Ovchinnikov: Dokl. AN USSR, 274(6), 1410 (1984) (in Russian).

[56] J.N. Aneli: Thesis of Doctor Dissertation, Georgian Technical University, Tbilisi, (1995) (in Russian).

[57] N.N. Kolossova, K.A. Boitsov: Physics of Solid State, 21(8), 2314 (1979) (in Russian).

[58] A.F. Tikhomirov, A.K. Pugachev, O.I. Olshevski, B.I. Sazhin: Plasticheskie Massi, 5, 13 (1998) (in Russian).

[59] V.G. Pavlii, A.E. Zaikin, E.V. Kuznetsov, L.N. Mikhailova: Izvestia Vuzov, Khimia i Tekhnologia, 6, 45 (1972) (in Russian).

[60] T. Slupkovski: Acta Polonica Physica, 148(2), 191 (1975).

[61] T. Slupkovski, R. Zielinski: Physika Status Solidi, 90(A), 737 (1985).

[62] M. Ghotraniha, I. Salovey: Polymer Eng. Sci., 28(1), 58 (1988).

[63] V.N. Gorshenev, B.A. Kamaritzki, V.M. Mikhailov, B.D. Saidov: Abstr. of Communications, Organic Materials for Electronics.International, Tashkent, 257 (1987) (in Russian).

[64] T.A. Ezquerra, M. Kubescza, F. Beita-Calicia: J. Synth. Metals, 41(3), 915 (1991).

[65] N.B. Pokrovskaya, A.A. Nikitin, B.N. Mayboroda: Khimicheskie Volokna, 4, 58 (1972) (in Russian).

[66] L.Z. Shenfil, L.V. Gerbova, N.A. Abramova, G.K. Melnikova, V.E. Gul': Kauchuk I Rezina, 7, 29 (1969) (in Russian).

[67] T. Lamond, C. Price: Rubber Age, 152(4), 49 (1970).

[68] A.E. Kornev, A.A. Blinov, V.S. Juravlev: Production of tires, Rubber Technology Articles and Asbotechnical Articles, 10, 5 (1969) (in Russian).

[69] 70. E.N. Ratnikov, Yu.L. Pogosov, G.L. Melnikova: Plasticheskie Massi, 1, 34 (1973) (in Russian).

[70] J. Meier: Polymer Eng. Sci, 13, 462 (1973).

[71] R. Gilg: Kunstoffberater, 22(5), 262 (1977).

[72] V.E. Gul', V.P. Sokolova, G.A. Klein S.Z. Bondarenko, E.A. Aripov, A.M. Berliand: Plasticheskie Massi, 10, 47 (1972) (in Russian).

[73] A.P. Losoto: Thesis of Candidate Dissertation, D.I. Mendeleev Chem. Tech. Institute, Moscow, (1982) (in Russian).

[74] J.YAKUBOVICH and M.NARKIS, Polymer Enginering and Science, 30(8)459(1990)

[75] A.E. Kornev, V.P. Kvardashov, V.F. Kormiushko, A.P. Zhukov: Production of Tires, Rubber Technical and Asbotechnical Articles, 5, 17 (1978) (in Russian).

[76] R.A. Gorelik, A.E. Kornev, A.V. Solomatin, A.T. Blok, V.S. Juravlev, A.I. Gorokhovskaya: Production of Tires, Rubber Technical and asbotechnical Articles, 10, 4 (1969) (in Russian).

[77] A.K. Sircar, T.G. Lamond: Rubber Chem.Technol., 51(1), 126 (1978).

[78] W.F. Verhelst, K.G. Wolthuis, A. Voet, P. Ehrburger, J.B. Donnet: Rubber Chem.Technol., 50(4), 735 (1977).

[79] G. Riess, J.B. Donnet: Rev.Gener.Caout., 41(3), 435 (1964).

[80] J.B. Donnet, J. Metzger: Rev.Gener.Caout., 41(3), 440 (1964).

[81] A.M. Gessler: Rubber Chem.Technol., 42(3), 585 (1969).

[82] K.E. Perepelkin, V.S. Smirnov, O.S. Karimarchik: Khimicheskie Volokna, 2, 6 (1980) (in Russian).

[83] E.N. Marmer: Carbon-Graphite Materials, Moscow, Metallurgia, 135 (1973) (in Russian).

[84] J.N. Aneli, D.I. Gventsadze, L.G. Shamanauri: Plasticheskie Massi, 1, 22 (1993) (in Russian).

[85] J.N. Aneli, D.I. Gventsadze, G.P.Mkheidze, L.G.Shamanauri: Reports on All-Union Conf. on Polymer Composites, Leningrad, 12 (1990) (in Russian).

[86] A.P. Vinogradov, A.K. Sarychev: Zh.Eksperim.Teor.Fiz., 85(3), 9 (1983) (in Russian).

[87] A.M. Lukyanova, E.N. Sofronova, A.G. Sharonova: Kauchuk i Rezina, 7, 26 (1983) (in Russian).

[88] V.M. Os'kin, A.E. Kornev: Chemistry and Technology of Rubber Processing, Leningrad, 88 (1989) (in Russian).

[89] Yu.P. Adler: Introduction into Experiment Planning, Moscow, Metallurgia, 157 (1969) (in Russian).

[90] J.N. Aneli, E.B. Vasil'eva, N.I. Rozova: Kauchuk i Rezina, 11, 20 (1988) (in Russian).

[91] W. Hofmann: Vulcanization and Vulcanizates, Hilfsmittel, Leverkusen, 464 (1968).

[92] Yu.S. Lipatov: Physical and Chemical Properties of Filled Polymers, Moscow, Khimia, 260 (1991) (in Russian).

In: Polymer Yearbook – 2011.
Editors: G. Zaikov, C. Sirghie et al. pp. 89-96

ISBN 978-1-61209-645-2
© 2011 Nova Science Publishers, Inc.

Chapter 8

ELECTRIC CONDUCTIVITY OF POLYMER COMPOSITES AT MECHANICAL RELAXATION

J. N. Aneli[*], G. E. Zaikov[**] and O. V. Mukbaniani[*]

[*]I.Javakhishvili Tbilisi State University,Chem.Department, Tbilisi, Georgia
[**]Emanuel Institute of Biophysical Physics of RAS, Moscow, Russia

ABSTRACT

Effect of mechanical relaxations on the relaxation phenomena of electrical conducting polymer composites (ECPC) have been investigated. It is experimentally shown that filler content significantly affects relaxation characteristics of the material. The main reason of these differences is the effect of the filler type on proceeding of relaxational processes. Rubbers with active carbon blacks display the ranges of slow and fast relaxations much more clearly, than those with low active carbon blacks even at increased concentrations in increase of the rubber elasticity modulus with the increase of the filler concentration. The effect of the filler type is displayed by formation of an inter-phase layer and is similar to the effect of the filler content. In the case of active carbon blacks, the increase of the filler-filler and polymer-filler interactions is balanced by high content of low active carbon blacks, because in both cases the modulus of the material and internal friction in relaxation processes grow.

Keywords: electrical conducting polymer composites; filler concentration; electrical conductivity; resistance; mechanical deformations; stretching; relaxation time.

1. INTRODUCTION

Mechanical stresses, induced by deformation of the composite materials, affect both the molecular and super-molecular structures of the polymers [1-3]. Structural changes occurring in composites produced by various mechanical deformations (stretching, compression, shear, etc.) affect the structure of an electrically conducting system. As a consequence, they affect the substructure and filler particles bound to it [4-6]. It is evident that in the case of conducting polymer composites mechanical deformation influences on the electrical conductivity of materials.

2. RESULTS AND DISCUSSION

It is well known that forces formed in polymer matrix from the very beginning of deformation counteracting the external influence. Therefore, the elucidation of relaxation effects on electric conductivity is complicated during deformation of ECPC due to overlapping of several factors.

To study electrically conducting properties of polymer composites during their mechanical relaxation the tests were performed, which used the same composites, studied at high deformations [7].

The experiment on determination of electrical volumetric specific resistance ρ during relaxation of the mechanical tension σ was performed as follows. A thin sheet (2 mm) sample of electrically conducting rubber based on polydimethylmethylvinilsiloxan (industrial type – SKTV) was stretched on a stretching machine at a definite deformation ϵ. The change of ρ was recorded by an automatic recorder immediately after stretching stop. The automatic recorder scale was graduated in Ohms on time τ, and synchronous recording of the time dependence of σ at fixed deformation was made. Moreover, relaxation characteristics were recorded for the same samples after the end the 'stretching-contraction' cycle.

As Figure1a shows, the values of ρ grow during relaxation of tension σ. In this case, the increase of ρ proceeds with higher rate in rubbers, deformed at relatively high rates. Moreover, increase of the stretching rate induces acceleration of ρ increase in these materials. In this case, kinetics of mechanical tension σ reduction is completely coincident with the character of ρ change. This means that the sample relaxes mechanically the faster, the higher the rate of the sample deformation is (Figure 1,b). Growth of ρ with time during relaxation of σ can be explained basing on the effect of the elastomer molecular system disordering that occurs after deformation release. In this case, internal mechanical tensions are reduced by transition of highly regulated system into lower regulated state on the present stage of deformation. Disordering in the system of macromolecules induces conducting circuit destruction, i.e. growth of ρ of the material. In view of the fact that the growth of the deformation rate induces increase of structural transformation intensities in the material, it is evident that significant deformations of the conducting system will happen in the electrically conducting system interacting with the polymer matrix.

The analysis of time dependences of ρ at the relaxation of rubbers show that they are successfully described by the expression:

$$\rho = \rho_\infty - \left(\rho_\infty - \rho_0\right)e^{-t/\tau},$$ (1)

where ρ_0 and ρ_∞ are border values of ρ during relaxation under fixed stretching deformation; τ is the relaxation time.

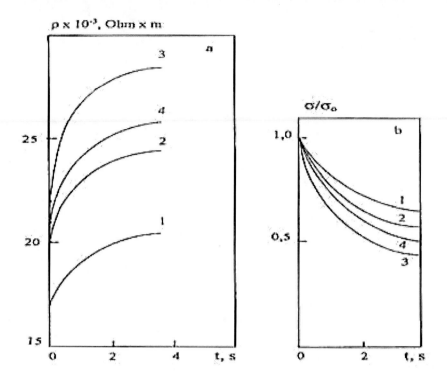

Figure 1. Time dependences of ρ (a) and σ (b) during relaxation of mechanical tension after stretching of the rubber, based on SCTV with P357E carbon black (350 mass parts) by 50 (1, 4), 100 (2), and 150% (3). The rate of elongation is 2.5 (1 - 3) and 5.0 mm/s (4), respectively

Eq. (1) represents the solution of the differential equation, similar to that [1] deduced for kinetics of the polymer mechanical relaxation. According to Eq. (1) kinetics of relaxation are described by the Maxwell model [2]:

$$\frac{d(\sigma - \sigma_\infty)}{dt} = E_1 \frac{dE}{dt} - \frac{\sigma_1 - \sigma_\infty}{\tau}.$$

Taking logarithm from Eq. (1), we get:

$$\ln \frac{\rho - \rho_\infty}{\rho_0 - \rho_\infty} = -\frac{t}{\tau}. \qquad (2)$$

Thus in the case, when application of Eq. (1) to relaxation kinetics is correct, the dependence should fit a straight line in coordinates of Eq. (2).

As Figure2 shows, time dependences by Eq. (2) are straightened. This allows to use the model of consequently connected strings and damper (the Maxwell model) for describing kinetics of the change of ρ during relaxation of mechanical tension of electrically conducting rubber.

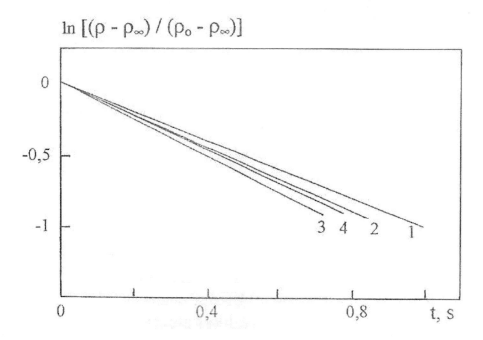

Figure 2. The dependence of ρ of the samples based on SCTV elastomer with P357E carbon black (50 mass parts) on time during relaxation in coordinates of Eq. (1). Numeration of curves coincides with one presented on Figure 1

Relaxation times τ for the composites of SCTV elastomer with P357E carbon black were determined by tangents of straight lines (Figure 2). Comparison of mechanical relaxation characteristics with analogical ones obtained for electrical conductivity of same materials leads to conclusion that the mechanical and electrical conductivity relaxations are in good correlation. Moreover by means of electrical conductivity there are possible to obtain in the solid polymers new relaxation phenomena invisible for other methods.

Table 1. Relaxation characteristics of stretched ECPC on the basis of SCTV*

Filler, mass parts	Deformation, %	Deformation rate, mm/s	Relaxation time τ, s
P357E (50)	50	2.5	1.00
	50	5.0	0.88
	100	2.5	0.91
	150	2.5	0.84
P357E (60)	30	2.5	0.60
	60	2.5	0.44
P803 (50)	100	2.5	3.84
P803 (60)	100	2.5	2.28

* Elastomers were obtained by the additive vulcanization technique.

The data shown in Table 1 testify that the increase of deformation of polymers leads to the growth of the relaxation rate of both ρ and σ. This effect is strengthened at the increase of the filler concentration in the composite. For example, at comparable rates of deformation ρ of rubber containing 60 mass parts of active carbon black relaxes with higher rate than in the same rubber containing 50 mass parts of the same carbon black. In this case, the effect of the interphase growth is observed. This growth causes an increase of conditionally equilibrium modulus of the system. In composites containing lower active carbon black (P803) relaxation times are much higher than in rubbers with active carbon blacks. This point out a relatively low adhesive interaction between the polymer and the filler in rubbers with P803 carbon black. However, weak absorption forces also grow at deformation (tension increases).

The investigation of relaxation processes at the end of the deformation cycle (stretching-contraction) after complete discharge of the samples showed that relaxation phenomena are more complex if compared with the above described facts. First of all, the complexity is expressed in the functional ρ-t dependence starting from the complete end of the cycle (Figure 3). Since the curves of the present figure were considered, let us mention that they, in fact, represent a superposition of at least two relaxation processes. One of these processes relates to the most linear parts of the curves, and the second one - to most curvilinear parts with limit overwhelming at $t \to \infty$ (see Figure 3, curves 2 and 3). Both parts of the curves mentioned reflect regulation of the conducting system by regulating the macromolecular part of the composite. However, the rate of relaxation and time of reaching equilibrium values of ρ at the end of the relaxation period significantly depend on the filler concentration. For the composite containing 40 mass parts of carbon black, the process proceeds at a low rate, but the equilibrium state in it occurs rather rapidly, whereas in rubbers with 50 and 60 mass parts of P357E higher rates are observed at the initial stage of relaxation, and decelerated curve growth - on the further stages of kinetics. One more fact is characteristic for the present dependences: the difference between minimum and maximum values of ρ on the whole interval of relaxation also strictly depends on the filling degree.

Analysis of the dependences, shown in Figure 3, gives a possibility to conclude the following. Taking into account the ideas by Bartenev /1/ on the formation of a complex heterogeneous system after injection of active carbon blacks into a polymer, which are characterized by non-linear viscoelasticity (the processes of this type are non-linear relaxation at low deformations and non-linear viscoelasticity at high deformation, the tixotropic effect by Mallins-Patrikeev [6], in particular), proceeding of two physical processes at relaxation of the considered systems can be supposed. The first process is close to the elastic range of deformation (Hooke's range) of a macromolecular system. As it is known, this process proceeds fast by both branches of the deformation cycle and is defined by elastic properties of the composite matrix. Elastic forces rapidly drives the system into the equilibrium after external influence release. However, it is often difficult to separate this process in polymers. In the composite material electrically conducting system, connected to the adsorption forces, can clearly response to any smallest structural changes in topology of structural polymer units (globules, for example), absolutely controlled by them. In the present case, conducting system in the polymer matrix plays the role of a relay-contact scheme translating information about the state of the surrounding (polymer) medium to the 'language' of electric conductivity. Thus, the abrupt decrease of ρ values, shown in Figure 3, corresponds to reduction of conducting channels existing in tight connection with macromolecules of elastomer, elastic properties of which is often promoted by the bonds mentioned.

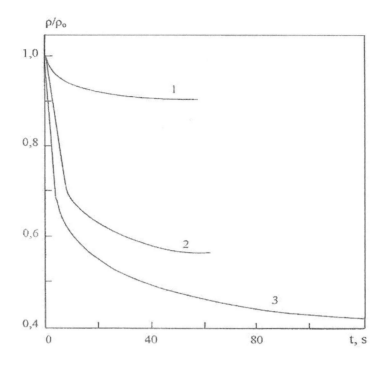

Figure 3. The dependence of ρ on delay after the contraction end (relaxation without loading) for rubbers, based on SCTV containing 40 (1), 50 (2) and 60 (3) mass parts of P357E carbon black, respectively. At the moment of discharge $\rho_0 = \rho$.

Table 2. Relaxation characteristics of electrically conducting rubbers based on SCTV elastomer at the end of deformation cycles*

No	Filler, mass parts	Sample prehistory		Relaxation time, s	
		Maximal deformation in the cycle, %	Deformation rate, %	Fast	Slow
1.	P357E (40)	150	2.5	1.2×10^{-2}	7
2.	P357 (50)	150	2.5	4.0	21
3.	P357E (60)	60	2.5	6.7×10^{-2}	28
4.	P803 (60)	200	2.5	—	6
5.	P803 (120)	150	2.5	2.2×10^{-2}	11
6.	P357 (40)	150	2.5	10.6×10^{-2}	48
7.	P803 (60)	150	2.5	5.1×10^{-2}	24
8.	P803 (60) + SCTN (20)**	200	2.5	2.8×10^{-2}	16

* Composites No. 1 - 5 were obtained by the additive vulcanization technique;
Composites No. 6 - 8 were obtained by the peroxide vulcanization technique.
** SCTN –low molecular (M.m. about 25 000) elastomer.

The second process is the reduction of initial system (before the deformation) with participation of filler particles, splitted from macromolecules during deformation (stretching) of the polymer system or absent in the adsorption connection with macromolecules during deformation (carbon black agglomerates, for example). As it is known, these particles decelerate regulation in the system of macromolecules, intensify internal friction and non-linear effects at high deformations, connected to it. That is why, their existence in the composite always leads to elongation of relaxation processes. Apparently, this is the reason of deceleration of restoration of the primary conducting system of the composite material. Thus, taking into account the above mentioned it becomes clear the delay of complete reduction of the equilibrium structure in high-filled polymer matrices in higher degree comparing with low-filled polymers. For the first glance, this position seems to be paradox. Let us remind now that according to the exponential dependence of ρ on concentration of the active filler reduction of the structure is much higher than for the composite possessing 50 mass parts of the same carbon black than for the composite with 60 mass parts of the same carbon black.

The ρ - τ dependences shown on Figure 3 reflect morphological changes in heterogeneous system. Analysis of these curves showed that exponential parts are successfully described by the (1) type of equation with the only difference that ρ_∞ is substituted by ρ_m at the end of relaxation, and ρ_0 by ρ_1, corresponded to the value of ρ at crossing exponential curve by the straight line.

Using Eq. (1) for various composites tested, relaxation times were determined. Numerical values of them are shown in Table 2.

The analysis of the data from Table 2 shows that:

1. Filler content significantly affects relaxation characteristics of the material. For example, the rubber containing 40 mass parts of P357E carbon black possesses the rate of fast relaxation about 5 times lower, and duration of slow relaxation is 4-fold lower than for analogous rubber with 60 mass parts of the same filler. The rubber containing 60 mass parts of P803 carbon black has no range of fast relaxation at all, but the same rubber displays such ranges at the filling degree of 120 mass parts of P803. The main reason of these differences is in increase of the rubber elasticity modulus with the increase of the filler concentration. On the one hand, this leads to the increase of the interphase surface square, and on the other hand, to occurrence of broken polymer-filler bonds, which promote growth of internal friction and deceleration of the relaxation process in the macromolecular system. That is why, although the increase of the modulus is the reason of growth of the rate of conditionally called fast relaxation, the mechanical tensions caused by accumulation of the filler particles, detached from polymer globules, leads to a noticeable increase of the slow relaxation time.

2. The effect of the filler type on proceeding of relaxation processes is also sufficient. Rubbers with active carbon blacks (P357E) display the ranges of slow and fast relaxations much more clearly, than those with low active carbon blacks (P803) even at increased concentrations. First of all, the effect of the filler type is displayed by formation of an interphase layer and is similar to the effect of the filler content. In the case of active carbon blacks, the increase of the filler-filler and polymer-filler

interactions is balanced by high content of low active carbon blacks, because in both cases the modulus of the material and internal friction in relaxation processes grow.

3. Comparison of characteristics of the rubber, produced by two different vulcanization techniques, shows that the rate and time of relaxation depends on vulcanization network density is not weaker than the type and concentration of the filler (compare composites 1 and 6, 4 and 7, Table 2). Since the concentration of polymer cross-links in peroxide vulcanizates significantly prevails over the concentration of longitudinal links (by SCTV end groups) in rubbers, obtained by additive vulcanization, it is evident that the mechanical modulus of the first composites is higher than that of the second ones. This is the reason of differences in relaxation characteristics. In this case, the increase of the 'soft phase' in the composite by introduction of low-molecular SCTN elastomer softens relaxation processes. This is expressed by decrease of fast relaxation rate and time of slow relaxation, if compared with similar composites containing no SCTN (compare composites 7 and 8, Table 2).

3. CONCLUSION

The measure of the mechanical relaxation processes in the electrical conducting polymer composites, particularly in conducting rubbers, may be investigated by study of time-dependent processes of electrical conductance, because the behaviour of the conducting system is directly depended on any changes in polymer matrix.

REFERENCES

[1] Bartenev G.M., Structure and relaxation properties of elastomers. Moscow: Khimia, 1979. 288 p. (In Russian).

[2] Casale A.,Porter R. Polymer stress reactions. New-York, Acad. Press, 1978, 264 p.

[3] Nielsen L. Mechanical properties of polymers and composites on their basis. Marcell Decker Inc, New-York, 1974. 276p.

[4] Donnet A.,Voet A. Carbon Black, Marcell Decker, New –York- Basel, 1976. -215 p.

[5] Carbon black polymer composites: the physics of electrically conducting composites. Marcell Decker, New-York, 1982. 246p.

[6] Aneli J.N., Khananashvili L.M., Zaikov G.E. Structuring and conductivity of polymer composites. New-York, Nova science publishers, 1998. 326 p.

[7] Aneli J.N. Zaikov G.E.., Khananashvili L.M. Effect of mechanical deformations on conductivity of conducting polymer composites. Journal of Applied Polymer Science, 1999, v.74, pp. 601-621.

In: Polymer Yearbook – 2011.   ISBN 978-1-61209-645-2
Editors: G. Zaikov, C. Sirghie et al. pp. 99-105   © 2011 Nova Science Publishers, Inc.

Chapter 9

DIFFUSION BEHAVIOUR OF TRIVALENT METAL IONS IN AQUEOUS SOLUTIONS

Ana C. F. Ribeiro, Artur J. M. Valente and Victor M. M. Lobo[1]

Department of Chemistry, University of Coimbra, 3004-535 Coimbra, Portugal

ABSTRACT

Trivalent metal ions have a relevant impact in different kinds of industry; for example, indium(III) and europium(III) ions are of practical importance in the development of new semiconductors and luminescent probes, respectively. Transport properties of ions and salts in aqueous solutions are important physical-chemical parameters allowing a better understanding of the behaviour of these ions in solution and so, helping to better describe the mechanism of processes taking place in their presence. However, the measurement of those transport properties is complicated due to the occurrence of hydrolysis; that may justify the scarcity of, e.g., diffusion data for aqueous solutions of europium(III) and indium(III) chlorides.

In this study, mutual diffusion coefficients for aqueous solutions of $InCl_3$ and $EuCl_3$, in a concentration range 0.002 mol dm^{-3} to 0.01 mol dm^{-3}, at 298.15 K, are reported. The open-ended conductometric capillary cell was used. The results are discussed on the basis of the Onsager-Fuoss and Pikal models.

Keywords: diffusion coefficients; europium(III); indium(III); aqueous solutions

1. INTRODUCTION

Europium(III) is a trivalent lanthanide ion with attractive and versatile spectroscopic and magnetic properties [1,2], which are an advantage for applications in different fields such as biochemistry or materials. Eu(III) can be used as a luminescent probe of bioactive species including metal ions, oxyanions and acidity of biological environments [3-5]. Eu(III) can also

---

[1] Corresponding author: vlobo@ci.uc.pt.

be applied for the study of surfactant association in solution [6]. The luminescent properties of Eu(III) have also been used for the development light-emitting diodes (LED) with an improved red emission [7,8]. Recently, europium-quantum dots and europium-fluorescein composite nanoparticles for the metal ion detection have been developed [9]. Eu(III) spectroscopy has also been used to characterize the complicated structural evolution that takes place during the gelation and densification of materials prepared by the sol-gel process [10].

Indium(III) chloride is an efficient catalyst for inducing various types of organic reactions, such as the synthesis of saccharides [11], Mukaiyama aldol reactions [12], Diels-Alder reactions [13,14], aza-Michael reactions [15], and also in microwave irradiation assisted synthesis [16]; other practical applications of indium(III) chloride includes, for example, its use as a constituent of a photosensitizer used as a photodynamic therapy agent for ocular diseases [17]. However, one of the most broad application of In(III) is as indium tin oxide (ITO), a solid solution with excellent electrical and optical properties [18]. In the last few years, the combined used of these metals has been attempted in order to improve the properties of LEDs, by using Eu(III) as a buffer layer on ITO [19], or introduce luminescent properties to ITO films or nanoparticles [20,21].

However, the use of Eu(III) and In(III) in solids and solutions requires an understanding of the factors affecting the properties of the ion. In this paper, we report diffusion coefficients of aqueous solutions Eu(III) and In(III) chlorides at 298.15 K, contributing for a better knowledge of the behavior those ion in solution.

2. EXPERIMENTAL

2.1. Reagents

The solutes used in this study were indium chloride and europium chloride (Aldrich, *pro analysi* > 97 % and *pro analysi* > 99.9 %, respectively) without further purification. Aqueous solutions were prepared using bi-distilled water. All solutions were freshly prepared just before each experiment.

2.2. Diffusion Measurements

The open-ended capillary cell used was constructed in this laboratory and is essentially the same as that previously reported [22]. The cell has two vertical capillaries, each closed at one end by a platinum electrode and positioned one above the other with the open ends separated by *ca.* 14 mm.

The upper (top) and lower (bottom) tubes, initially filled with solutions of concentrations $0.75c$ and $1.25c$, respectively, were surrounded with a solution of concentration c. This ambient solution was contained in a $200 \times 140 \times 60$ mm glass tank, which was immersed in a bath thermostatted at 298.15 K. The tank was divided internally by Perspex sheets, while a glass stirrer created a slow lateral flow of ambient solution across the open ends of the capillaries. Experimental conditions were such that the concentration at each of the open ends was equal to the ambient solution value c, that is the physical length of the capillary tube coincided with the diffusion path, such that the boundary conditions described in [22] to solve Fick's second law of diffusion are applicable. Therefore, the so-called Δl-effect [22] is reduced to negligible proportions. In contrast to a manual apparatus, where diffusion is followed by measuring the ratio of resistances of the top and bottom tubes, w = R_t/R_b, by an alternating current transformer bridge, in our automatic apparatus w was measured by a Solartron digital voltmeter (DVM) 7061 with 6 1/2 digits. A Bradley Electronics Model 232 power source supplied 30 V (stable to ±0.1 mV) to a potential divider that applied a 250 mV signal to the platinum electrodes in the top and bottom capillaries. By rapidly (< 1 s) measuring the voltages V' and V'' from the top and bottom electrodes relative to the central electrode at ground potential the w was then calculated from the DVM readings.

To measure the differential diffusion coefficient D at a given concentration c, 2 dm^3 each of a "top" solution of concentration $0.75c$ and a "bottom" solution $1.25c$ were prepared. The "bulk" solution of concentration c was produced by mixing accurately measured volumes of 1 dm^3 of "top" solution with 1 dm^3 of "bottom" solution. The glass tank and the two capillaries were filled with solution c, immersed in the thermostat, and were allowed to come to thermal equilibrium. The quantity $TR_{inf} = 10^4/(1 + w)$ was now measured very accurately (where w = R_t/R_b is the electrical resistance ratio for solutions of concentration c of the top (t) and bottom (b) diffusion capillaries at infinite time). $TR = 10^4/(1 + w)$, is the equivalent, at any time t.

The capillaries were then filled with "top" and "bottom" solutions, which were allowed to diffuse into the "bulk" solution. Resistance ratio readings were taken at various times, beginning 1000 min after the start of an experiment. The diffusion coefficient was evaluated using a linear least-squares procedure to fit the data, followed by an iterative process which uses 20 terms of the expansion series of the solution of Fick's second law for the present boundary conditions. The theory developed for this cell has been described previously [22].

Table 1. Diffusion coefficients, $\overline{D}$, of EuCl$_3$ in aqueous solutions at different concentrations, c at 298.15 K

c/ /mol dm^{-3}	$\overline{D}$ [a]/ /10^{-9} m^2 s^{-1}	$S_{\overline{D}}$ [b]/ /10^{-9} m^2 s^{-1}	D_{OF} [c]/ /10^{-9} m^2 s^{-1} (a = 5.6×10^{-10} m [d])	D_{Pik} [c]/ /10^{-9} m^2 s^{-1} (a = 5.6×10^{-10} m [d])	$\Delta D/D_{OF}$ [e] / %	$\Delta D/D_{Pik}$ [e] / %
2×10^{-3}	1.216	0.010	1.180	1.519	3.0	−19.9
3×10^{-3}	1.200	0.013	1.169	1.795	2.6	−33.1
5×10^{-3}	1.179	0.020	1.158	2.205	1.8	−46.5
8×10^{-3}	1.160	0.011	1.153	3.009	0.6	−61.4
1×10^{-2}	1.151	0.010	1.152	3.994	−0.1	−71.2

a $\overline{D}$ is the mean diffusion coefficient of 3 experiments. b $S_{\overline{D}}$ is the standard deviation of that mean. cDOF and DPik represent the calculated diffusion coefficients from Onsager-Fuoss and Pikal equations, respectively. dSum of hydrated ionic radii (diffraction methods) [23]. eΔD/DOF and ΔD/DPik represent the relative deviations between $\overline{D}$ and DOF and DPikal values, respectively.

3. RESULTS

Tables 1 and 2 show the experimental diffusion coefficients, D, of EuCl3 and InCl3 in aqueous solutions at 298.15 K. These results are the average of 3 experiments performed on consecutive days. The experimental procedure shows good reproducibility, as shown by the small standard deviations, SDav. The accuracy of the systems (uncertainty 1-2 %) has been demonstrated by measurements on other solutions of different electrolytes (e.g., [2-16]).

Table 2. Diffusion coefficients, $\overline{D}$, of InCl$_3$ in aqueous solutions at different concentrations, c at 298.15 K

c/ /mol dm^{-3}	$\overline{D}$ [a]/ /10^{-9} m^2 s^{-1}	$S_{\overline{D}}$ [b]/ /10^{-9} m^2 s^{-1}	D_{OF} [c]/ /10^{-9} m^2 s^{-1} (a = 5.3 × 10^{-10} m [d])	D_{Pik} [c]/ /10^{-9} m^2 s^{-1} (a = 5.3 × 10^{-10} m [d])	$\Delta D/D_{OF}$ [e] / %	$\Delta D/D_{Pik}$ [e] / %
2×10^{-3}	0.945	0.021	1.081	0.921	−12.6	2.6
3×10^{-3}	0.867	0.026	1.077	0.844	−19.5	2.7
5×10^{-3}	0.686	0.023	1.072	0.698	−36.0	−1.7
8×10^{-3}	0.622	0.020	1.067	0.477	−41.2	30.3
1×10^{-2}	0.619	0.011	1.067	0.370	−42.0	67.3

[a] $\overline{D}$ is the mean diffusion coefficient of 3 experiments. [b] $S_{\overline{D}}$ is the standard deviation of that mean. [c] D_{OF} and D_{Pik} represent the calculated diffusion coefficients from Onsager-Fuoss and Pikal equations, respectively. [d] Sum of hydrated ionic radii (diffraction methods) [23]. [e] $\Delta D/D_{OF}$ and $\Delta D/D_{Pik}$ represent the relative deviations between $\overline{D}$ and D_{OF} and D_{Pikal} values, respectively.

The following polynomial in c was fitted to the data by a least squares procedure,

$$D = a_0 + a_1 c + a_2 c^2 \qquad (1)$$

where the coefficients a_0, a_1, and a_2 are adjustable parameters. Table 3 shows the coefficients a_0 to a_2 of eq. (1). They may be used to calculate values of diffusion coefficients at specified concentrations within the range of the experimental data shown in Tables 1 and 2. The goodness of the fit (obtained with a confidence interval of 98 %) can be assessed by the correlation coefficient, R^2.

Table 3. Fitting coefficients (a_0-a_2) of a polynomial equation [$D/(10^{-9}$ m^2 s^{-1})$ = a_0 + a_1$ (c/mol dm^{-3}) + a_2 (c/mol dm^{-3})]^2 to the mutual differential diffusion coefficients of europium chloride and indium chloride in aqueous solutions at 298.15 K

Electrolyte	a_0	a_1	a_2	R^2
EuCl$_3$	1.246	-17.06	766.7	0.997
InCl$_3$	1.196	-139.4	8228	0.990

4. DISCUSSION

4.1. Limiting diffusion coefficients

Extrapolation of the fit of these equations to infinitesimal concentration gives the estimated diffusion coefficients obtained (*i.e.*, D^0 = a$_0$ in Table 3), which account for the diffusion of both the cation and the anion under these conditions. As can be seen in Table 4 the agreement between these values and those obtained by Nernst equation [24,25] (eq. 2), using different values for equivalent conductance of europium and indium at infinitesimal concentration, is reasonable.

$$D^0 = \frac{RT}{F^2} \frac{|z_1| + |z_2|}{|z_1 z_2|} \frac{\lambda_1^0 \lambda_2^0}{\lambda_1^0 + \lambda_2^0} \qquad (2)$$

λ_1^0 and λ_2^0 represent the equivalent conductance of the cation and anion at infinitesimal concentration, respectively, and z_1 represents the algebraic valency of a cation and z_2 is the algebraic valency of an anion.

Table 4. Limiting diffusion coefficients, D^0, for the systems EuCl$_3$/H$_2$O and InCl$_3$/H$_2$O

Electrolyte	D^0_{exp} [a]/ /(10^{-9} m^2 s^{-1})	D^0_{Nernst}/ /(10^{-9} m^2 s^{-1})	$\Delta D^0/D$ %[d]
EuCl$_3$	1.246	1.235[b]	0.9
		1.274[b]	−2.2
InCl$_3$	1.196	1.150[c]	4.0

[a] Limiting D^0_{exp} values were calculated by extrapolating our experimental data, D_{exp} (Tables 1 and 2) to c → 0 at 298.15 K. [b] Diffusion coefficients estimated by Nernst' equation (eq. 2), using λ (Eu^{3+}) = 192 × 10^{-4} S m^2 mol^{-1} and λ (Eu^{3+}) = 203.4 × 10^{-4} S m^2 mol^{-1}, respectively. These values were obtained by using a Stokes-Einstein equation and experimental values, respectively [26]. [c] Diffusion coefficients estimated by Nernst' equation (eq. 2), using the value λ (In^{3+}) = 168.9 × 10^{-4} S m^2 mol^{-1} obtained by Campbell et al. [27].

4.2. Inter Ionic Effects on Diffusion

Having in mind to understand the transport process of this electrolyte in aqueous solutions, the experimental mutual diffusion coefficients at 298.15 K were compared, as a first approach, with those estimated by the Onsager-Fuoss and Pikal equations (eqs. 5 and 10 [24,28,29]) (Tables 1 and 2).

The first equation is expressed by

$$D = \left(1 + c\frac{\partial \ln y_\pm}{\partial c}\right)\left(D^0 + \Sigma \Delta_n\right)$$

(3)

where D is the mutual diffusion coefficient of the electrolyte, the first term in parenthesis is the activity factor, $y_\pm$ is the mean molar activity coefficient, c is the concentration in mol dm^{-3}, D^0 is the Nernst limiting value of the diffusion coefficient (eq. 2), and Δ_n are the electrophoretic terms given by

$$\Delta_n = k_B T\, A_n \frac{\left(z_1^n t_2^0 + z_2^n t_1^0\right)^2}{|z_1 z_2| a^n}$$

(4)

where k_B is the Boltzmann's constant; T is the absolute temperature; A_n are functions of the dielectric constant, of the viscosity of the solvent, of the temperature, and of the dimensionless concentration-dependent quantity (ka), being k the reciprocal of average radius of the ionic atmosphere; t_1^0 and t_2^0 are the limiting transport numbers of the cation and anion, respectively.

Since the expression for the electrophoretic effect has been derived on the basis of the expansion of the exponential Boltzmann function, because that function is consistent with the Poisson equation, we only would have to take into account the electrophoretic term of the first and the second order (n = 1 and n = 2). Thus, the experimental data D$_{exp}$ can be compared with the calculated D$_{OF}$ on the basis of eq. (5)

$$D = \left(1 + c\frac{\partial \ln y_\pm}{\partial c}\right)\left(D^0 + \Delta_1 + \Delta_2\right)$$

(5)

The theory of mutual diffusion in binary electrolytes, developed by Pikal [28], includes the Onsager-Fuoss, but has new terms resulting from the application of the Boltzmann exponential function for the study of diffusion. On the other words, instead of approximating the Boltzmann exponential by a truncated power series, the calculations are performed retaining the full Boltzmann exponential. As a result of this procedure, a term representing the effect of ion-pair formation appears in the theory as a natural consequence of the electrostatic interactions. The electrophoretic correction appears now as the sum of two terms

$$\Delta v_j = \Delta v_j^L + \Delta v_j^s$$

(6)

where Δv_j^L represents the effect of electrostatic interactions of long-range, and Δv_j^s represents them as short- range.

Designating by $M = 10^{12}\ L/c$ the solute thermodynamic mobility, where L is the thermodynamic diffusion coefficient, ΔM can be represented by the equation

$$\frac{1}{M} = \frac{1}{M^0}\left(1 - \frac{\Delta M}{M^0}\right) \tag{7}$$

where M^0 is the value of M for infinitesimal concentration, and

$$\Delta M = \Delta M^{OF} + \Delta M_1 + \Delta M_2 + \Delta M_A + \Delta M_{H1} + \Delta M_{H2} + \Delta M_{H3} \tag{8}$$

The first term on the right hand in the above equation, ΔM^{OF}, represents the Onsager-Fuoss term for the effect of the concentration in the solute thermodynamic mobility, M; the second term, ΔM_1, is a consequence of the approximation applied on the ionic thermodynamic force; the other terms result from the Boltzmann exponential function.

The relation between the solute thermodynamic mobility and the mutual diffusion coefficient is given by

$$D = \frac{L}{c}10^3 RTv\left(1 + c\frac{\partial ln\, y_\pm}{\partial c}\right) \tag{9}$$

where R is the gas constant, and v is the number of ions formed upon complete ionization of one solute "molecule". From equations (7) and (9), we obtain a version of Pikal's equation more useful for estimating the mutual diffusion coefficients of electrolytes, D_{Pikal}. That is,

$$D\,Pikal = \frac{10^3 RTv}{\dfrac{1}{M^0}\left(1 - \dfrac{\Delta M}{M^0}\right)}\left(1 + c\frac{\partial ln\, y_\pm}{\partial c}\right) \tag{10}$$

Both Onsager-Fuoss and Pikal's theories introduce the ion size parameter a, distance of closest approach from the Debye-Huckel and it is well known that there is no direct method for measuring this parameter. In this work, the values for a were estimated from Marcus data (Table XIII of Ref. 23), using two approximations (Table 5). Firstly, the a-values were estimated as the sum of the ionic radii (R_{ion}) reported by Marcus [23]. The R_{ion} values were obtained as the difference between the mean internuclear distance of a monoatomic ion, or the central atoms of polyatomic ions, and the oxygen atom of a water molecule in its first hydration shells ($d_{Ion\text{-}water}$), and the half of the mean intermolecular distance between two water molecules in liquid water (R_{water}). Briefly,

$$R_{ion} = d_{ion\text{-}water} - R_{water} \text{ and } a = R_{cation} + R_{anion}$$

In order to account for the effect of the ion hydration shell on the a-values, a second approximation considers the sum of the $d_{\text{ion-water}}$ values reported by Markus [23]. In other words, in this approach the a-values are determined as $a = R_{\text{cation-water}} + R_{\text{anion-water}}$ (Table 5).

Table 5. Values of mean distances of closest approach, $a/10^{-10}$ m, of the systems $EuCl_3/H_2O$ and $InCl_3/H_2O$ calculated by two methods

Electrolyte	Sum of ionic radii in solutions [a] $a = R_{\text{cation}} + R_{\text{anion}}$	Sum of mean ion-water internuclear distances [a] $a = d_{\text{cation-water}} + d_{\text{anion-water}}$
$EuCl_3$	2.86	5.64
$InCl_3$	2.56	5.34

[a]See ref. 23.

For $EuCl_3$, we see agreement between experimental data and Pikal calculations is not good (Table 1), eventually because of the full use of Boltzmann's exponential in Pikal's development. However, Onsager-Fuoss theory leads to calculated values close to the experimental data (deviations ≤ 3 %, Table 1). In these estimations, the choice of the parameter a was irrelevant, within reasonable limits. We may use any value because slight variations in this parameter a have little effect on the final results of D_{OF} and D_{Pikal}. For $InCl_3$ (Table 2), Pikal's treatment gives better agreement with D_{exp} than Onsager-Fuoss for dilute solutions. However, in this case, the final result D_{Pikal} is strongly affected by the choice of this parameter a. In fact, there is a better agreement with D_{Pikal} if we use a value of parameter a equal to the sum of the hydrated ionic radii. Despite the limitations of this theory when applied to non symmetrical electrolytes, this good applicability in this case can lead us to admit that the effects of short range interactions on the diffusion of this electrolyte at those concentrations are relevant, contrary in the $EuCl_3$. The hydration of this salt, its hydrolysis [30] and the eventual formation of ion pairs, increasing with concentration, can be responsible for those effects.

For c > 0.008 M, the results predicted from the above model differ markedly from experimental observation (i.e., 30 % - 67 %). This is not surprising if we take into account the change with concentration of parameters such as viscosity [25, 31, 32], dielectric constant [25] and hydration [25, 31-33], which are not taken into account in these models.

REFERENCES

[1]   Lanthanide Probes in Life, Chemical and Earth Sciences: Theory and Practice. J.C.G. Bünzli, G.R. Choppin, Eds.; Elsevier: Amsterdam, 1989.
[2]   M.J. Lochhead, P.R. Wamsley, K.L. Bray. *Inorg. Chem.* 1994, 33, 2000-2003.
[3]   R. J. P. Williams. Struct. Bonding 1982, 50, 79-119.
[4]   D. Costa, M.L. Ramos, H.D. Burrows, M.J. Tapia, M.G. Miguel. *Chem. Phys.* 2008, 352, 241-248.

[5]   E.J. New, D. Parker, D.G. Smith, J.W. Walton. *Curr. Opin. Chem. Biol.* 2010, 14, 238-246.

[6]   A.J.M. Valente, H.D. Burrows, R.F. Pereira, A.C.F. Ribeiro, J.L.G. Costa Pereira, V.M.M. Lobo. Langmuir 2006, 22, 5625–5629.

[7]   Y.H. Zhou, L. Zhou, J. Wu, H.Y. Li, Y.X. Zheng, X.Z. You, H.J. Zhang. Thin Solid Films 2010, 518, 4403-4407.

[8]   P. He, H.H. Wang, S.G. Liu, J.X. Shi, G. Wang, M.L. Gong. Electrochem. Solid State Lett. 2009, 12, 61-64.

[9]   H.T. Dong, Y. Liu, D.D. Wang, W.Z. Zhang, Z.Q. Ye, G.L. Wang, J.L. Yuan. Nanotechnology 2010, 21, 395504.

[10] M.J. Lochhead, K.L. Bray. *J. Non-Cryst. Solids* 1994, 170, 143-154.

[11] D. Mukherjee; P.K. Ray, U.S. Chowdhury. Tetrahedron 2001, 57, 7701-7704.

[12] T.P. Loh, J. Pei, G.Q. Cao, *Chem. Commun.* 1996, 1819.

[13] J. Zhang, C.J. Li. *J. Org. Chem.* 2002, 67, 3969-3971.

[14] M.L. Kantam, M. Roy, S. Roy, M.S. Subhas, B. Sreedhar, B.M. Choudary, R.L. De. J. Mol. Catal. A-Chem. 2007, 265, 244-249.

[15] a) T.P. Loh, L.L. Wei, *Synlett* 1998, 975-976. b) J. Zhang, C.J. Li. *J. Org. Chem.* 2002, 67, 3969-3971.

[16] N.L. Chavan, P.N. Naik, S.K. Nayak, R.S. Kusurkar. Synth. Commun. 2010, 40, 2941-2947.

[17] T.A. Ciulla, M.H. Criswell, W.J. Snyder, W. Small. Br. *J. Ophthalmol.* 2005, 89, 113-119.

[18] C.G. Granqvist, A. Hultaker. Thin Solid Films 2002, 411, 1-5.

[19] S.W. Shi, D.G. Ma. Phys. Status Solidi A-Appl. Mat. 2009, 206, 2641-2644.

[20] J.K. Kim, Y.G. Choi. Thin Solid Films 2009, 517, 5084-5086.

[21] D.P. Dutta, V. Sudarsan, P. Srinivasu, A. Vinu, A.K. Tyagi. J. Phys. Chem. C 2008, 112, 6781-6785.

[22] J.N. Agar, V.M.M. Lobo. *J. Chem. Soc., Faraday Trans.* I 1975, 71, 1659-1666.

[23] Y. Marcus. Chem. Rev. 1988, 88, 1475-1498.

[24] R. A. Robinson, R. H. Stokes. Electrolyte Solutions, 2nd Ed., Butterworths: London, 1959.

[25] H.S. Harned, B.B. Owen. The Physical Chemistry of Electrolytic Solutions, 3rd Ed., Reinhold Pub. Corp.: New York, 1964.

[26] E. Mauerhofer, K. Zhernosekov, F. Rosch, Radiochim. Acta 2003, 91, 473-477.

[27] A.N. Campbell, *Can. J. Chem.* 1973, 51, 3006-3009.

[28] M.J. Pikal, *J. Phys. Chem.* 1971, 75, 663-675.

[29] V.M.M. Lobo, A.C.F. Ribeiro and S.G.C.S Andrade. *Ber. Buns. Phys. Chem.* 1995, 99, 713-720.

[30] C.F. Baes Jr., R.E. Mesmer. The Hydrolysis of Cations, John Wiley and Sons: New York, 1976, pp. 358-365.

[31] A.R. Gordon, *J. Chem. Phys.* 1937, 5, 522-526.

[32] V.M.M. Lobo, A.C.F. Ribeiro. *Corros. Prot. Mat.* 1994, 13, 18-23.

[33] V.M.M. Lobo, A.C. F. Ribeiro. Port. Electrochim. Acta 1995, 13, 41-62.

In: Polymer Yearbook – 2011.   ISBN 978-1-61209-645-2
Editors: G. Zaikov, C. Sirghie et al. pp. 107-117   © 2011 Nova Science Publishers, Inc.

Chapter 10

POTASSIUM AND SODIUM 2,6-DI-TERT-BUTYL PHENOXIDES AND THEIR PROPERTIES

A. A. Volod'kin and G. E. Zaikov

N.M. Emanuel Institute of Biochemical Physics, Russian Academy of Sciences[1],
Moscow 119334, Russia

ABSTRACT

The determining factor of the reaction of 2,6-di-tert-butylphenol with alkaline metal hydroxides is temperature, depending on which two types of potassium or sodium 2,6-di-tert-butyl phenoxides are formed with different catalytic activity in the alkylation of 2,6-di-tert-butylphenol with methyl acrylate. More active forms of $2,6\text{-}Bu^t_2C_6H_3OK$ or $2,6\text{-}Bu^t_2C_6H_3ONa$ are synthesized at temperatures higher than 160°C represent predominantly monomers of 2,6-di-tert-butylphenoxides producing dimmers on cooling. The data of NMR ^{1}H, electronic, and IR spectra for the corresponding forms of $2,6\text{-}Bu^t_2C_6H_3OK$ and $2,6\text{-}Bu^t_2C_6H_3ONa$ isolated in the individual state showed a cyclohexadienone structure. In DMSO or DMF media, the dimeric forms of 2,6-di-tert-butylphenoxides react with methyl acrylate to form methyl 3-(4-hydroxy-3,5-di-tert-butylphenyl) propionate in 64—92% yield.

Keywords: phenols, phenoxides, 2,6-di-tert-butylphenol, methyl acrylate, Michael reaction, kinetics, dimers, sodium hydroxide, potassium hydroxide.

Compounds based on 2,6-di-tert-butylphenol were multiply used as objects of studies for development of theoretical concepts of the chemistry of sterically hindered phenols and for synthesis of 4-substituted phenols. Among the known reactions, those involving 2,6-di-tert-butyl phenoxides, which are reactants and catalysts in syntheses of substances with useful properties, are poorly studied. Information on the chemistry and technology of alkylation of $2,6\text{-}But2C6H3OH$ with methyl acrylate in the presence of potassium or sodium 2,6-di-tert-butyl phenoxides is predominantly concentrated in patents [1-5] from which it follows that $2,6\text{-}But2C6H3OK$ and $2,6\text{-}But2C6H3ONa$ are synthesized by the reactions of 2,6-

---

[1] Kosygin 4 str., Moscow 119334, Russia,chembio@sky.chph.ras.ru.

But2C6H3OH with alkaline metal alkoxides or hydroxides at elevated temperature in vacuo. It is also known that at room temperature 2,6-But2C6H3OH does not react with alkaline metal hydroxides, [6] whereas at 100—140oC the rate of this process is insufficient [7] for the preparation of 2,6-di-tert-butyl phenoxides in the individual form due to reversibility of the process. The studied reactions of 2,6-dialkyl phenoxides are described in the review [8] and publications [9-13] associated with the use of the corresponding phenoxides in the catalytic alkylation of 2,6-dialkylphenols. Despite interest in the alkylation of 2,6-But2C6H3OH, the data on the properties of individual phenoxides are insufficient, scanty, and contradictory. It remains unclear what compounds are formed in the reactions of 2,6-But2C6H3OH with alkaline metal alkoxides or hydroxides, and no data on the structure, properties, and reactivity of individual alkaline metal 2,6-di-tert-butylphenoxides are available. In the patent [5] 2,6-di-tert-butylphenoxides are described as monomers, whereas other literature sources [14, 15] show that some derivatives of 2,6-di-tert-butylphenoxides exist as dimers. Therefore, it seems of interest to reveal the true structure of 2,6-$But_2C_6H_3OK$ and 2,6-$But_2C_6H_3ONa$ and optimize the method of their synthesis. In the present work we continued previous studies [7] of the reactions of 2,6-$But_2C_6H_3OH$ with potassium or sodium hydroxides at different temperatures and found that with the temperature increase from 160 to $200^{\circ}C$ the reaction rate becomes sufficient for the formation of steam as bubbles in the reaction mixture. As a result of water evaporation, the equilibrium shifts to the formation of alkaline metal phenoxide. Under these conditions, 2,6-$But_2C_6H_3OK$ and 2,6-$But_2C_6H_3ONa$ are formed as crystals, which made it possible to isolate them in the individual state and study their properties. It turned out that the catalytic activities differ substantially for 2,6-$But_2C_6H_3OK$ synthesized by the reaction of 2,6-$But_2C_6H_3OH$ with KOH at 180—$200^{\circ}C$ and the material obtained from 2,6-$But_2C_6H_3OH$ and $But OK$ in $But OH$ when the solvent is removed in vacuo ~$100^{\circ}C$. The process at 100—$140^{\circ}C$ produces a uniform gel-like mixture consisting of 2,6-di-tert-butylphenol, the corresponding phenoxide, and alkaline metal hydroxide.

EXPERIMENTAL

NMR ^1H spectra were recorded on a Bruker WM-400 instrument (400 MHz) in DMSO-d_6 (99.5%) sing signals of the solvent as internal standard. The integral intensity from the water peak was less than 5% of the integral intensity of signals of residual protons of DMSO-d_6. IR spectra were obtained on a Perkin—Elmer 1725-X spectrometer in KBr pellets (resolution 2 cm^{-1}). Electronic spectra were measured on a Shimadzu UV-3101 PC instrument. Molecular weights were determined by the thermoelectric method using a cell described previously[16] and calculated by the formula Equation,

$$M = (k_g M_s / 100(t_1 - t_2)$$

where g is a weighed sample (g) per 100 g of solvent (2—3%), k is the coefficient of the instrument and thermistors, M_s is the molecular weight of a standard sample, and $t_1 - t_2$ is the value equivalent to the temperature change in the cell filled with a sample solution by the potentiometer scale. The temperature in the thermostat was 82°C. Kinetic data were obtained by liquid chromatography of reaction mixtures (Bruker LC-31 chromatograph, IBM Cyano column, hexane—isopropyl alcohol ethyl acetate (8 : 1 : 1, v/v) mixture as eluent, rate 0.4 mL min^{-1}).

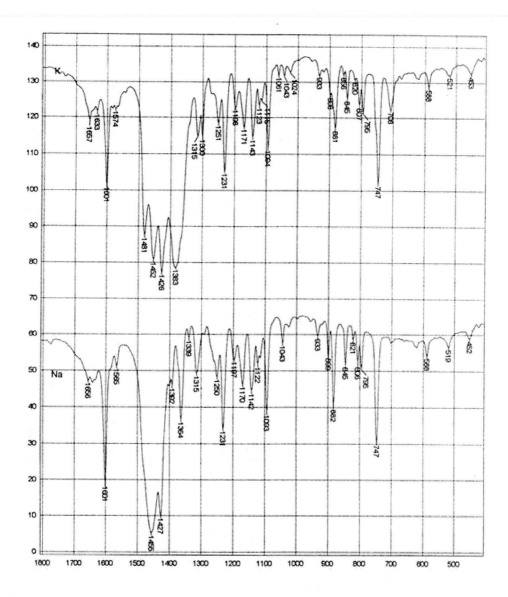

Transmittance / Wavenumber (cm-1)   Stacked Y-Zoom CURSOR

Figure 1. IR spectra of the monomer (a) and dimer (b) of potassium 2,6-di-tert-butyl phenoxide synthesized by methods A and B, respectively.

Potassium 2,6-di-tert-butyl Phenoxide

A. A flask was filled with 2,6-Bu$^t_2$C$_6$H$_3$OH (20.6 g, 0.1 mol) and heated in an argon flow to 190°C, and granulated (85%) KOH (1.5 g) was added. After 10 minutes, the crystals that formed were separated by high-temperature filtration under argon and washed with n-octane heated to 110—115°C until no 2,6-Bu$^t_2$C$_6$H$_3$OH was found in the mother liquor. Residues of the solvent were separated in vacuo at 120—130°C to obtain 2,6-Bu$^t_2$C$_6$H$_3$OK (5.3 g, 93.5%), which was placed under argon into a flask heated to 120—130°C. NMR ^{1}H, δ: 1.30 (s, 18 H); 5.58 (t, 1 H, J = 6 Hz); 6.58 (d, 2 H, J = 6 Hz). NMR ^{1}H spectrum of 2,6-di-tert-butylphenol (for comparison) , δ: 1.37 (s, 18 H); 3.48 (s, 1 H); 6.74 (t, 1 H, J = 8 Hz); 7.06 (d, 2 H, J = 8 Hz). IR, ν/cm^{-1}: 1657, 1633 (—C=C—C=O, monomer); 1601 (—C=C—C=O, dimer); 1574 (—C=C—, aromatic structure) (Figure 1, a). Found: M$_{min}$ = 260. Calculated for 2,6-Bu$^t_2$C$_6$H$_3$OK: M = 244.28. Upon heating the crystals above 220°C, they are partially decomposed with release of isobutylene. When this product is treated with 10% HCl, the LC analysis shows the presence of 2-tert-butylphenol.

B. A flask was filled with 2,6-Bu$^t_2$C$_6$H$_3$OH (20.6 g, 0.1 mol) and heated in an argon flow to 190°C, and granulated (85%) KOH (1.5 g) was added. After 10 minutes, the reaction mixture as a suspension was cooled to ~20°C, and heptane (50 mL) was added. The precipitate was filtered off and washed with heptane until no 2,6-Bu$^t_2$C$_6$H$_3$OH was found in the washing solutions to obtain 5.1 g (90%) of 2,6-Bu$^t_2$C$_6$H$_3$OK.

Its NMR ^{1}H spectrum is identical to that recorded for the sample synthesized using method A. IR, ν/cm^{-1}: 1657, 1633 (—C=C—C=O, monomer); 1601 (—C=C—C=O, dimer); 1574 (—C=C—, aromatic structure) (Figure 1, b). UV, λ$_{max}$/nm (log□): 320 (3.48) (solvent DMSO); 317 (3.72) (solvent DMF). Found (%): C, 68.54; H, 8.40; K, 16.25. C$_{14}$H$_{21}$KO. Calculated (%): C, 68.79; H, 8.66; K, 16.00. Found: M$_{max}$ = 480. Calculated for 2,6-Bu$^t_2$C$_6$H$_3$OK: M = 244.28. On cooling the 2,6-Bu$^t_2$C$_6$H$_3$OK sample synthesized by method A, the intensity of the bands in the IR spectrum at 1700—1300 cm^{-1} changes below ~20°C, and the spectrum becomes identical to that recorded for the sample obtained using method B.

Sodium 2,6-di-tert-butylphenoxide

Sodium 2,6-di-tert-butylphenoxide was synthesized similarly to method B in 86% yield. NMR ^{1}H, δ: 1.30 (s, 18 H); 5.59 (t, 1 H, J = 6 Hz); 6.57 (d, 2 H, J = 6 Hz). IR, ν/cm^{-1}: 1656, 1633 (—C=C—C=O, monomer); 1601 (C=C—C=O, dimer); 1565 (—C=C—, aromatic structure). UV, λ$_{max}$/nm (logε): 320 (3.55) (solvent DMSO); 322 (3.48) (solvent DMF). Found (%):C, 73.54; H, 9.40; Na, 10.25. C$_{14}$H$_{21}$NaO. Calculated (%): C, 73.65; H, 9.28; Na, 10.07. Found: M$_{max}$ = 458. Calculated for 2,6-Bu$^t_2$C$_6$H$_3$ONa: M = 228.96.

Methyl 3-(4-hydroxy-3,5-di-tert-butylphenyl) Propionate

A. Methyl acrylate (2.5 mL, 0.03 mol) was added in an argon flow to a solution of 2,6-Bu$^t_2$C$_6$H$_3$OK (4.88 g, 0.01 mol), which was synthesized by method B, in DMSO (9.4 mL) at 115°C. After three hours, the reaction mixture was cooled to ~20°C, 10% HCl was added to neutral pH, and the product was extracted with hexane. Methyl 3-(4-hydroxy-3,5-di-tert-butylphenyl) propionate was obtained in a yield of 5.16 g (88%), m.p. 66°C (cf. Ref. 9: m.p. 66°C). When DMSO was replaced with DMF, the yield of the product became 92%. When 2,6-Bu$^t_2$C$_6$H$_3$ONa in an equivalent amount (4.56 g) was used instead of 2,6-Bu$^t_2$C$_6$H$_3$OK in DMSO, the product yield was 64% (LC data).

B. Granulated KOH (0.11 g, 0.002 mol) was added in an argon flow to 2,6-Bu$^t_2$C$_6$H$_3$OH (20.6 g, 0.1 mol) heated to 190°C. After 10 minutes, the reaction mixture was cooled to 130°C, and methyl acrylate (11 g, 0.13 mol) was added. After the reactants were mixed, the temperature of the reaction mixture decreased to 110°C. After 25 minutes, the content of the product in the reaction mixture was 98 mol.%.

C. Methyl acrylate (11 g, 0.13 mol) was added in an argon flow at 110°C to a mixture of 2,6- Bu$^t_2$C$_6$H$_3$ OH (20.6 g, 0.1 mol) and 2,6-Bu$^t_2$C$_6$H$_3$OK (0.488 g, 0.002 mol) synthesized by method B. After 25 minutes, the content of the product in the reaction mixture became 15 mol.%, while after three hours, it reached 87%. Similarly, when 2,6-Bu$^t_2$C$_6$H$_3$ONa (0.456 g, 0.002 mol) synthesized by method B was used, the product yield after three hours became 72%.

RESULTS AND DISCUSSION

2,6-Di-tert-butylphenol does not react with alkaline metal hydroxides at room temperature, whereas on heating to 140°C the acid-base equilibrium is achieved with a low rate and shifts to the initial components [7]. However, at a higher temperature (180—200°C) the reaction rate is sufficient for steam bubbles to form in a solution of 2,6-di-tert-butylphenol, which favors their removal from the reaction mixture and the shift of equilibrium. The reaction proceeds within a short time (5—10 min) and is accompanied by the formation of potassium or sodium 2,6-di-tert-butyl phenoxide as crystals. The study of 2,6-Bu$^t_2$C$_6$H$_3$OK and 2,6-Bu$^t_2$C$_6$H$_3$ONa isolated in the individual state found that they can exist as both monomers and dimers. At high temperature a monomer is formed and spontaneously transformed into a more stable dimeric form with the temperature decrease. The conclusion about the existence of two forms of the corresponding phenoxides is based on the results of measurements of the molecular weights, IR spectroscopic data, and observation of different catalytic activities in the reaction of 2,6-Bu$^t_2$C$_6$H$_3$OH with methyl acrylate. To retain the primary properties of the monomeric forms of 2,6-Bu$^t_2$C$_6$H$_3$OK or 2,6-Bu$^t_2$C$_6$H$_3$ONa phenoxides from the moment of formation to the corresponding measurements, we proposed a method for conservation of a sample by storage in a flask under argon at 120—130°C. In the IR spectra of the samples obtained by methods A and B (monomer and dimer, respectively), the most informative is the region from 1700 to 1300 cm^{-1} containing the characteristic bands at 1633 and 1656 cm^{-1} (typical of the conjugated carbonyl group of quinolide compounds) and at 1601 cm^{-1}. It is known [17] that the band at 1601 cm^{-1} is present in the IR spectrum of the dimer of 4-bromo-2,6-di-tert-butyl-4-methylcyclohexa-2,5-dienone, and it has earlier been assigned to the carbonyl group of the quinobromide compound in the dimer structure. In the IR spectra of the considered samples, the bands at 1633, 1656 cm^{-1} and 1601 cm^{-1} differ in intensity. For the 2,6-Bu$^t_2$C$_6$H$_3$OK monomer the intensity of the bands at 1633, 1655 cm^{-1} is much higher, whereas in the spectra of 2,6-Bu$^t_2$C$_6$H$_3$OK the high-intensity band lies at 1601 cm^{-1}. The band at 1574 cm^{-1}, which belongs to the C=C bond of aromatic compounds [18] has a low intensity in the spectra. The frequency region of 1475—1300 cm^{-1} usually remains uninterpreted; however, it contains bands for the most part of quinolide compounds and methylene quinones. The NMR ^{1}H spectra of potassium and sodium 2,6-di-tert-butyl phenoxides in DMSO-d$_6$ show that signals from the meta- and para-protons lie at δ 6.58 and 5.58, respectively, which corresponds to the region of olefinic protons. Compared to the spectrum of the initial 2,6-di-tert-butylphenol, these signals exhibit noticeable upfield shift. This result confirms that molecules of potassium and sodium 2,6-di-tert-butyl phenoxides in solution exist in the quinolide form. The shape of the electronic spectra of dimers of potassium and sodium 2,6-di-tert-butyl phenoxides in DMSO changes in time, which indicates, probably, the interaction of phenoxides with a polar aprotic solvent. The absorbance of the bands begins to change from the moment of preparation of the solution and continues during 4—6 h. This is accompanied by a decrease in the absorbance of the band with λ_{max} = 320 nm and an increase in the absorbances of the bands with λ_{max} = 260 and 481 nm, and these changes pass through the isosbestic points (Figure 2).

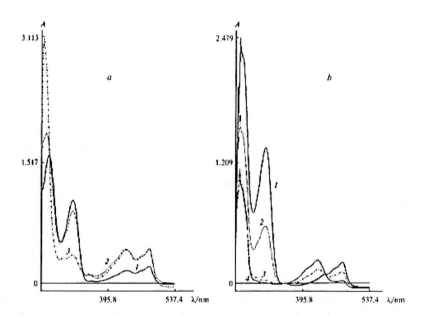

Figure 2. Changes in the electronic spectra of potassium 2,6-di-tert-butyl phenoxide in solutions of DMSO (a) and DMF (b); a: initial spectrum (1), spectrum in 48 (2) and 352 min (3) after dissolution; b: initial spectrum (1), spectrum in 37 (2), 118 (3), and 181 min (4) after dissolution.

It can be assumed that in a DMSO solution the 2,6-Bu$^t_2$C$_6$H$_3$OK dimer (band with λ_{max} = 320 nm) decomposes to form an ambidentate ion (λ_{max} = 260 nm) and a complex with the solvent (λ_{max} = 481 nm). The broad band with λ_{max} = 767 nm is present in the initial spectrum of 2,6-Bu$^t_2$C$_6$H$_3$OK and retained during transformation of the electronic spectrum with time. This fact can be related to vibrations of the metal cation in a molecule of the 2,6-Bu$^t_2$C$_6$H$_3$OK dimer or 2,6-Bu$^t_2$C$_6$H$_3$OK complex with the solvent. Their amplitude is sufficient for the formation of the electronic spectrum at $\lambda = 767$ nm. The time dependence of the absorbance of the band with $\lambda_{max} = 320$—317 nm is linear and characterized by the k_1 constant depending on the solvent nature. Similar results were obtained for the electronic spectra of sodium 2,6-di-tert-butylphenoxide in DMSO and DMF solutions. The initial spectrum of the 2,6-Bu$^t_2$C$_6$H$_3$ONa dimer contains bands with $\lambda_{max} = 320$ nm (DMSO) and 322 nm (DMF), which undergo transformations at 478—481 nm. Taking into account the obtained results, it seemed of interest to study regularities of the Michael reactions between the 2,6-Bu$^t_2$C$_6$H$_3$OK and 2,6-Bu$^t_2$C$_6$H$_3$ONa dimers and methyl acrylate in polar solvents, for instance, DMSO and DMF. It turned out that in these solvents the main direction is the reaction of alkaline metal 2,6-di-tert-butyl phenoxides with methyl acrylate affording methyl 3-(4-hydroxy-3,5-di-tert-butylphenyl)propionate. In the absence of solvent, the main direction is methyl acrylate polymerization [11]. The kinetics of the reaction of the 2,6-Bu$^t_2$C$_6$H$_3$OK dimer with methyl acrylate was studied in DMSO and DMF solutions. Comparing the experimental and calculated data for the kinetic scheme with account for the decomposition of the 2,6-Bu$^t_2$C$_6$H$_3$OK dimer in polar solvents, we calculated the apparent rate constants (k_2) of the reaction of the ArOK—Solv complexes (Solv is solvent) with methyl acrylate (Scheme 1).

$$(2,6\text{-}Bu^t_2C_6H_3OK) + Solv \underset{k_{-1}}{\overset{k_1}{\rightleftharpoons}} 2\,(2,6\text{-}Bu^t_2C_6H_3OK)\text{--}Solv$$

$$2\,(2,6\text{-}Bu^t_2C_6H_3OK)\text{--}Solv + MA \underset{k_{-2}}{\overset{k_2}{\rightleftharpoons}} 2,6\text{-}Bu^t_2C_6H_3OK\text{---}MA + Solv$$

$$2,6\text{-}Bu^t_2C_6H_3OK\text{-----}MA \overset{k_3}{\longrightarrow} Ar\,OK$$

$$Ar\,OK + H^+ \longrightarrow Ar\,OH$$

MA is methyl acrylate , ArOH is   methyl-3-(4hydroxy-3,5-di-tert-butylphenyl)propionate

Using Scheme 1, we performed the mathematical simulation of the experimental kinetic data for ArOK accumulation taking into account $k_1 = 7.5 \cdot 10^{-5}$ s^{-1}, which was determined from the plot of the decrease in the absorbance of the band with $\lambda_{max} = 320$ nm of the 2,6-Bu$^t_2$C$_6$H$_3$OK dimer in DMSO. The program for calculation of the reaction kinetics based on the solution of the "rigid system" of differential equations was used in mathematical simulation. The experimental data and calculated curve are shown in Figure 3.

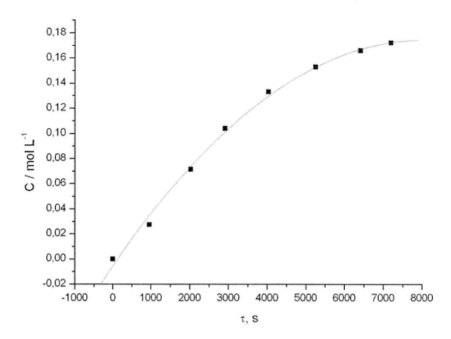

Figure 3. Experimental data (points) and results of calculation (curve) of the kinetics of formation of potassium methyl 3-(4-hydroxy-3,5-di-tert-butylphenyl) propionate in the reaction of the potassium 2,6-di-tert-butyl phenoxide dimer with methyl acrylate in a DMSO solution; C is the concentration of potassium methyl 3-(4-hydroxy-3,5-di-tert-butylphenyl) propionate; $[(2,6\text{-Bu}^t_2\text{C}_6\text{H}_3\text{OK})_2]_0 = 0.01$ mol L^{-1}, $[\text{MA}]_0 = 0.3$ mol L^{-1}, $[\text{DMSO}]_0 = 12$ mol L^{-1}, 115 °C.

The results of calculation of the parameters (k_1/s^{-1}, k_2/L mol^{-1} s^{-1}, k_{-2}/L mol^{-1} s^{-1}, k_3) for $[2,6\text{-Bu}^t_2\text{C}_6\text{H}_3\text{OK}]_2$ (I), MA (II), and solvent (III) are given below.

Solvent	C_0/mol L^{-1}			$k_1 \cdot 10^{-4}$	$k_2 \cdot 10^{-3}$	$k_{-2} \cdot 10^{-5}$	$k_3 \cdot 10^{-3}$
	I	II	III				
DMSO	0.1	0.3	12	0.75	2.3	4	6
DMF	0.1	0.3	12	3.8	6.3	4	2

The calculation of the kinetic scheme of the reaction of methyl acrylate with 2,6-Bu$^t_2$C$_6$H$_3$ONa and DMF gives the following reaction rate constants: $k_1 = 9.1 \cdot 10^{-4}$ s^{-1}, $k_2 = 9.7 \cdot 10^{-4}$, $k_{-2} = 7.4 \cdot 10^{-4}$ L mol^{-1} s^{-1}, and $k_3 = 8.2 \cdot 10^{-3}$. Alkaline metal 2,6-di-tert-butyl phenoxides can add to methyl acrylate as stoichiometric reactants. It was of interest to use them as catalysts of the reactions of 2,6-Bu$^t_2$C$_6$H$_3$OH with methyl acrylate. It turned out that the catalytic properties of 2,6-di-tert-butyl phenoxides in this reaction depend on the method of their preparation (method A or B). The 2,6-Bu$^t_2$C$_6$H$_3$OK dimer (method B) is similar in catalytic activity to the product of the reaction of 2,6-di-tert-butylphenol with potassium tert-butoxide, whereas the 2,6-Bu$^t_2$C$_6$H$_3$OK monomer (method A) is most efficient of the earlier known catalysts. When the content of the 2,6-Bu$^t_2$C$_6$H$_3$OK monomer is 1.5—3.0 mol.% of the content of 2,6-Bu$^t_2$C$_6$H$_3$OH at 110—115°C, the reaction ceases within 15—20 min to form ArOH in up to 98% yield (Figure 4, a).

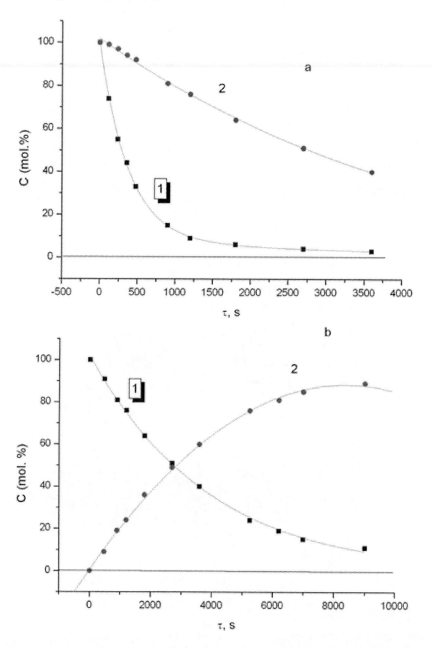

Figure 4. Kinetics of consumption of 2,6-Bu$^t_2$C$_6$H$_3$OH in the catalytic reaction with methyl acrylate in the presence of the monomer and dimer of 2,6-Bu$^t_2$C$_6$H$_3$OK (a) and 2,6-Bu$^t_2$C$_6$H$_3$OK synthesized from 2,6-Bu$^t_2$C$_6$H$_3$OH and Bu^{t}OK (b); a: 1, monomer; [ArOH]$_0$ = 3.29 mol L^{-1}, [MA]$_0$ = 3.75 mol L^{-1}, [ArOK]$_0$ = 0.066 mol L^{-1}, 115 °C; 2, dimer; the (2,6-Bu$^t_2$C$_6$H$_3$OK)$_2$ dimer synthesized from 2,6-Bu$^t_2$C$_6$H$_3$OH and ÊÎÍ at 190 °C and isolated at 20 °C, temperature of alkylation at 115 °C; [ArOH]$_0$ = 3.29 mol L^{-1}, [MA]$_0$ = 3.75 mol L^{-1}, [(ArOK)$_2$]$_0$ = 0.05 mol L^{-1}; b: 1, kinetics of consumption of 2,6-Bu$^t_2$C$_6$H$_3$OH; 2, kinetics of formation of methyl 3-(4-hydroxy-3,5-di-tert-butylphenyl) propionate.

Under similar conditions, the reaction of potassium 2,6-di-tert-butyl phenoxide synthesized in situ from 2,6-Bu$^t_2$C$_6$H$_3$OH and potassium tert-butoxide (>5% mol.%) proceeds within 2.5—3 h to form ArOH in a yield of at most 85% (Figure 4, b). The catalytic properties of the monomeric forms of 2,6-Bu$^t_2$C$_6$H$_3$ONa and 2,6-Bu$^t_2$C$_6$H$_3$OK are similar, whereas the reaction of 2,6-Bu$^t_2$C$_6$H$_3$OH with methyl acrylate in the presence of the 2,6-Bu$^t_2$C$_6$H$_3$ONa dimers proceeds with a lower rate.

Thus, we found that the catalytic properties of potassium (sodium) 2,6-di-tert-butyl phenoxides depend on the temperature of their synthesis by the reaction of 2,6-di-tert-butylphenol with alkaline metal hydroxides. Methyl 3-(4-hydroxy-3,5-di-tert-butylphenyl) propionate is formed predominantly in the reaction of the 2,6-Bu$^t_2$C$_6$H$_3$OK or 2,6-Bu$^t_2$C$_6$H$_3$ONa dimer with methyl acrylate in DMSO and DMF solutions. The data obtained indicate, possibly, that charge separation with electron density distribution between the O atom and C atoms of the cyclohexadienone structure occurs in a molecule of potassium and sodium 2,6-di-tert-butyl phenoxides.

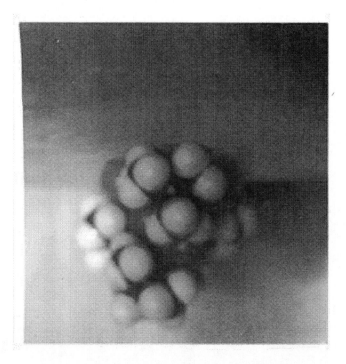

Figure 5. Atomic model of the alkaline metal 2,6-di-tert-butyl phenoxide dimer.

A consequence of polarization of a molecule of the phenoxides is their capability of dimerizing and interacting with metal cations through double bonding with the O and C atoms of the six-membered ring involving two phenoxide molecules. As follows from the atomic model of the dimer (Figure 5), the oxygen and metal atoms are spatially shieldedby the tert-butyl substituents, and their intramolecular arrangement favors the interaction of the metal cations with the O and C atoms of two six-membered rings. The monomeric form of

2,6-Bu$^t_2$C$_6$H$_3$OK or 2,6-Bu$^t_2$C$_6$H$_3$ONa phenoxides is presented by the nonassociated, more reactive metal cation, which just predetermines their higher catalytic properties.

REFERENCES

[1]   US Pat. 3277148 (1966); *Chem. Abstrs*, 1966, 65, 18535.

[2]   US Pat. 3526668 (1970); *Chem. Abstrs,* 1970, 73, 98589.

[3]   Jpn Pat. 161350 (1981); *Chem. Abstrs,* 1982, 96, 162344.

[4]   US Pat. 5264612 (1993); *Chem. Abstrs,* 1993, 120, 191356.

[5]   US Pat. 5177247 (1993); *Chem. Abstrs,* 1993, 116, 6249.

[6]   G. Stillson, *J. Am. Chem. Soc.*, 1946, 68, 722.

[7]   A.A. Volod´kin and G.E. Zaikov, Izv. Akad. Nauk, Ser. Khim., 2002, 2031 [*Russ. Chem. Bull., Int. Ed.,* 2002, 51, 2189].

[8]   A.A. Volod´kin and G.E. Zaikov, Ros. Khim. Zh., 2000, 44, No. 2, 81 [*Mendeleev Chem. J.,* 2000, 44, No. 2 (Engl. Transl.)].

[9]   A.A. Volod´kin, V. I. Paramonov, F. M. Egidis, and L. K. Popov, Khim. prom-st´ [Chemical Industry], 1988, No. 12, 7 (in Russian).

[10] A.A. Volod´kin, Izv. Akad. Nauk, Ser. Khim., 1994, 827 [Russ. Chem. Bull., 1994, 43, 769 (Engl. Transl.)].

[11] A.A. Volod´kin, A.S. Zaitsev, V.L. Rubailo, V.A. Belyakov, and G.E. Zaikov, Izv. Akad. Nauk SSSR, Ser. Khim., 1989, 1829 [Bull. Acad. Sci. USSR, *Div. Chem. Sci.,* 1989, 38, 1677 (Engl. Transl.)].

[12] T. F. Titova, A. P. Krysin, V. A. Bulgakov, and V. I. Mamatyuk, Zh. Org. Khim,, 1984, 20, 1899 [*J. Org. Chem.* USSR, 1984, 20 (Engl. Transl.)].

[13] V. A. Bulgakov, N. N. Gorodetskaya, G. A. Nikiforov, and V. V. Ershov, Izv. Akad. Nauk SSSR, Ser. Khim., 1983, 71 [Bull. Acad. Sci. USSR, *Div. Chem. Sci.,* 1983, 32 (Engl. Transl.)].

[14] K. Shobatake and K. Nakamoto, *J. Inorg. Chim.* Acta, 1970, 4, 485.

[15] E. Muller, P. Ziemek, and A. Rieker, Tetrahedron Lett., 1964, No. 4, 207.

[16] E. Yu. Bekhli, D. D. Novikov, and S. G. Entelis, Vysokomolek. Soedin., Ser. A, 1967, 9, 2754 [*Polym. Sci., Ser.* A, 1967, 9 (Engl. Transl.)].

[17] N.A. Malysheva, A.I. Prokof´ev, N.N. Bubnov, S.P. Solodovnikov, T.I. Prokof´eva, A.A. Volod´kin, and V. V. Ershov, Izv. Akad. Nauk SSSR, Ser. Khim., 1977, 1522 [Bull. Acad.Sci. USSR, Div. *Chem. Sci.,* 1977, 26 (Engl. Transl.)].

[18] L. PaQuette and W. Farley, *J. Org. Chem.,* 1967, 32, 2718.

In: Polymer Yearbook – 2011.
Editors: G. Zaikov, C. Sirghie et al. pp. 119-130

ISBN 978-1-61209-645-2

Chapter 11

REACTIONS OF DIETHYL N-ACETYLAMINO (3,5-DI-TERT-BUTYL-4-YDROXYBENZYL) MALONATE IN ALKALINE SOLUTION

A. A. Volod'kin, G. E. Zaikov and L. N. Kurkovskaia

N.M. Emanuel Institute of Biochemical Physics Russian Academy of Sciences[1]
Moscow 119334, Russia

ABSTRACT

Alkaline hydrolysis of diethyl *N*-acetylamino(3,5-di-*tert*-butyl-4-hydroxybenzyl) malonate is accompanied by decarboxylation. The efficiency of this process depends on the temperature and ratio of the reactants. A possibility of tautomerism with migration of the proton of phenolic hydroxyl and the influence of the structure on the antioxidation properties were considered on the basis of analysis of the IR spectral data and quantum chemical (PM6) calculation of the structures. The energies of homolysis of the OH bond of phenolic hydroxyl were calculated for a series of the synthesized compounds. It is proposed to predict the antioxidation activity on the basis of these values.

Keywords: N-acetylamino(3,5-di-tert-butyl-4-hydroxybenzyl)malonate, hydrolysis, decarboxylation, antioxidants.

INTRODUCTION

The use of malonic acid derivatives in the synthesis of tyrosine analogs with tert-butyl substituents in the aromatic ring made it possible to consider this method as optimal for the preparation of water-soluble antioxidants promising in biology and medicine. It is known [1] that the formylamide group in a molecule of 2-N-formylamino-3-(3,5-di-tert-butyl-4-

---

[1] 4 Kosygin str., Moscow 119334, Russia, chembio@sky.chph.ras.ru.

hydroxyphenyl)propanoic acid is deformylated by aniline to form 2-amino-3-(3,5-di-tert-butyl-4-hydroxyphenyl)propanoic acid. Alkaline hydrolysis of diethyl N-acetylamino(3,5-di-tert-butyl-4-hydroxybenzyl)malonate in a water-alcoholic solution gives the corresponding acid in 63% yield. The thermolysis of the acid at 140—180 □C produces 2-N-acetylamino-3-(3,5-di-tert-butyl-4-hydroxyphenyl)propanoic acid. It is insufficient to identify the earlier synthesized compounds by the 1H NMR spectra on the basis of multiplet signals from the protons of the —CH2—CH— fragment, because this identification does not allow one to determine the vicinal and geminal constants. The single example for the alkaline hydrolysis of this diethyl ester does not elucidate the regularities of this reaction and its synthetic potentialities. Therefore, the further study of alkaline hydrolysis reactions and properties of the compounds formed seems topical.

In the present work, we found that decarboxylation occurs simultaneously with the alkaline hydrolysis of diethyl N-acetylamino(3,5-di-tert-butyl-4-hydroxybenzyl)malonate, due to which the N-acetyltyrosine derivatives are formed along with the derivatives of acetylaminomalonic acid. The intermediate reaction products are corresponding water-soluble sodium or potassium salts

RESULTS AND DISCUSSION

The alkaline hydrolysis of diethyl N-acetylamino(3,5-di-$tert$-butyl-4-hydroxy-benzyl)malonate (**1**) results in the saponification of ester groups followed by the decarboxylation of the substituted malonic acid formed (Scheme 1). Under the of this malonic acid. The acidification of these salts results in the corresponding substituted malonic acid, whose properties differed from those described earlier [1] conditions of higher decarboxylation rates, the tyrosine analogs are formed, whose yield depends on the reactant ratio and reaction conditions. In a solution of aqueous dioxane at the mole ratio **1**:NaOH equal to 1:6 and temperature 100 °C monosodium salt **2a**, sodium salt of tyrosine **3à**, and sodium salt of monoester **2b** are formed in the process of conjugated reactions. The rate constants for the conjugated reactions (Scheme 2) were determined from the data of changing the composition of the reaction mixtures in time and by the calculation of the kinetic scheme using modeling of the system of differential equations (Gear program [2]).

The kinetic scheme is presented by 12 reactions with the initial concentrations of compound **1**, NaOH, and H_2O, whose change affects the results of the consumption of the initial compounds and accumulation of the intermediate and final compounds. At the ratio of the initial molar concentrations of compound **1** and NaOH equal to 1:6, the corresponding salts of carboxylic acids are formed, and the evolved CO_2 reacts with alkali, decreasing the concentration of NaOH. The kinetic scheme is presented by three blocks that describe the reactions of saponification and decarboxylation and the exchange reactions involving NaOH, H_2O, and CO_2.

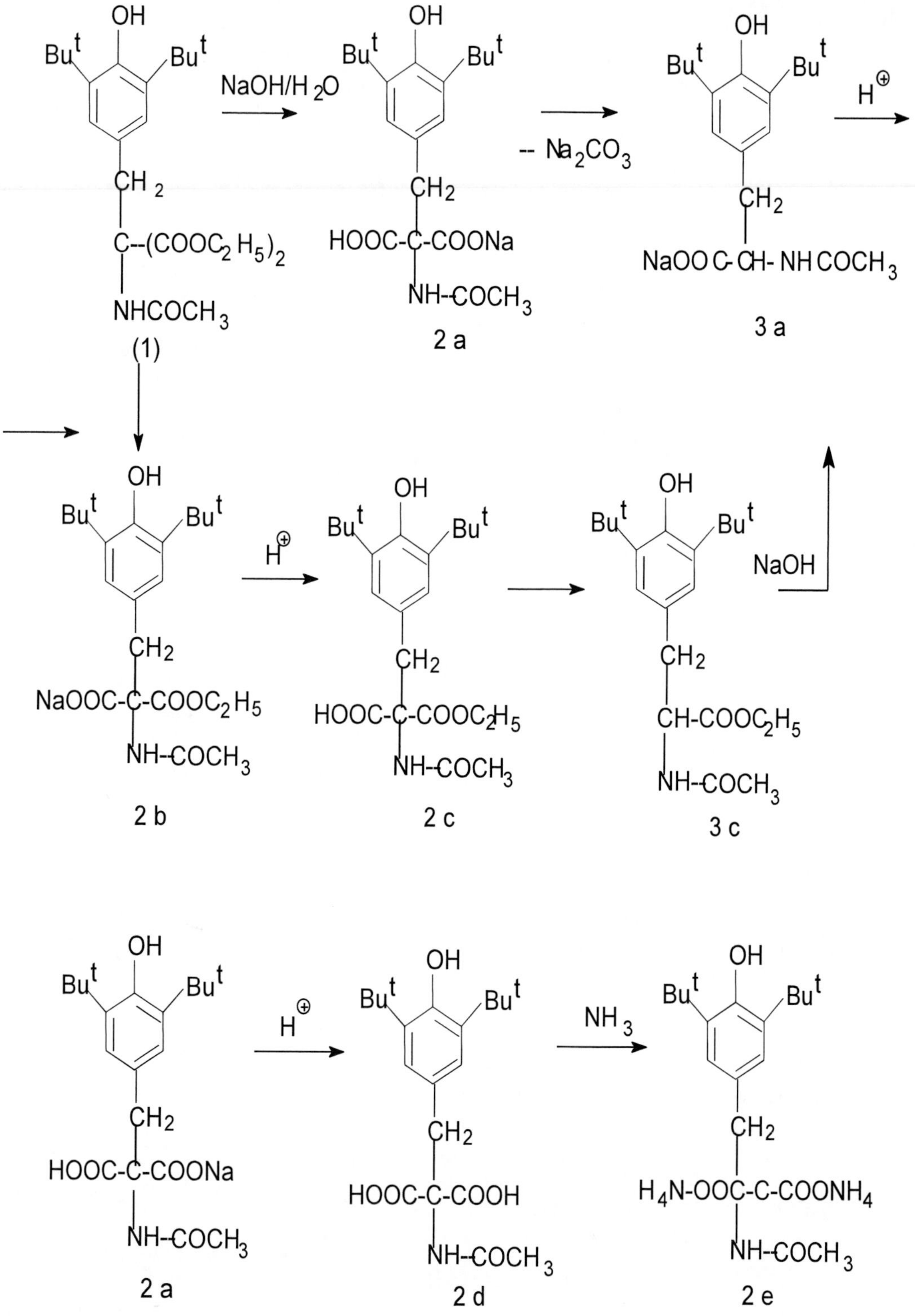
NaOH/H2O
-- Na2CO3
H⊕
(1)
2 a
3 a
H⊕
NaOH
2 b
2 c
3 c
H⊕
NH3
2 a
2 d
2 e

Kinetic scheme

№	reaction	k
1	1 + NaOH → 2b + EtOH	2.10-3L.mol-1.s-1
2	1 + NaOH + H2O → 2a + EtOH + EtOH	10–4 L2•mol2•s–1
3	1 + H2O + H2O → 2d + EtOH + EtOH	8.5•10–5 L2•mol2•s–1
4	2b + H2O → 2c + NaOH	1 L•mol–1•s–1
5	2c + NaOH → 2b + H2O	10 L•mol–1•s–1
6	2a + H2O → 2d + NaOH	1 L•mol–1•s–1
7	2d + NaOH → 2a + H2O	1•10–2 L•mol–1•s–1
8	2c + H2O → 3b + CO2 +EtOH	3.1•10–3 L•mol–1•s–1
9	2a + NaOH + NaOH → 3a + Na2CO3 +H2O	3•10–2 L•mol–1•s–1
10	2d → 3b + CO2	1.5•10–2 s–1
11	NaOH + CO2 → NaHCO3	1•102 L•mol–1•s–1
12	NaHCO3 + NaOH → Na2CO3 + H2O	1•102 L•mol–1•s–1

Figure 1 presents the experimental data on the consumption of compound **1** and the accumulation of the reaction products (after neutralization to pH 4). Figure 2 shows the results of calculation of the kinetics of this reaction. Satisfactory *coincidenc*e of the experiment and calculation allowed us to use the kinetic scheme in the development of the method of synthesis of 2a.

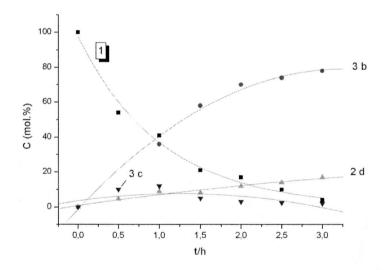

Figure 1. Experimental data on the consumption of compound 1 and accumulation of 2d, 3b, and 3c under the conditions of hydrolysis of compound 1 in aqueous dioxane at 100 □C. [1]0 = 0.357 mol L–1; [NaOH]0 = 2.24 mol L–1; [H2O]0 = 5.95 mol L–1.

It follows from the calculation of the kinetic scheme of the initial concentrations of compound **1** and NaOH equal to 1:1 that more than 50% of product **2a** are formed after 1.5—2 h (Figure 3). In the preparative experiment, compound **2a** was obtained in 62% yield.

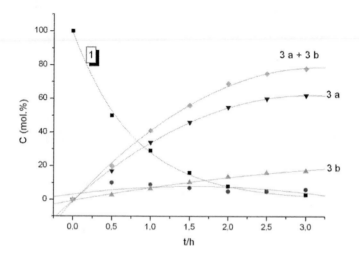

Figure 2. Calculated data for the kinetics of the conjugated reactions under the hydrolysis conditions in aqueous dioxane at 100 °C; $[1]0 = 0.357$ mol L–1; $[NaOH]0 = 2.24$ mol L–1; $[H2O]0 = 5.95$ mol L–1.

The hydrolysis conditions at room temperature in the presence of AcONa were developed for the preparation of compound **2c** in the individual form. On heating compound **2c** 175—180 °C decarboxylation occurs to form compound **3c**.

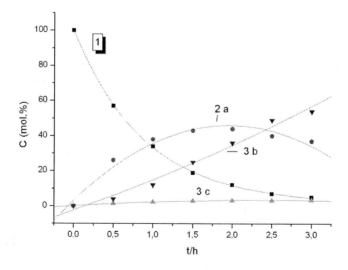

Figure 3. Calculated data for the kinetics of the conjugated reactions under the conditions of hydrolysis of compound 1 in aqueous dioxane at 100 °C; $[1]0 = 0.455$ mol L–1; $[NaOH]0 = 0.455$ mol L–1; $[H2O]0 = 5.05$ mol L–1.

The IR and ^{1}H NMR spectral data present sufficient information on the structure and properties of the obtained compounds. It is known [3] that the NH group in α-acetylamino acids appears in the region of 3390—3260 cm^{-1}, and vibrations of the Ñ-H bond in the acetylamide group are observed at 1620—1640 cm^{-1}. These characteristic bands are observed in the IR spectra of the synthesized compounds and, hence, no reactions occur at the acetylamide group under the experimental conditions. The corresponding signals in the ^{1}H NMR spectra of the synthesized compounds confirm the presence of the COCH$_3$ group (δ_i 1.91—1.75) and NH group (δ_i 7.69—7.05). The ^{1}H NMR spectra of the tyrosine derivatives indicate the presence of two nonequivalent hydrogen atoms in the CH$_2$ group, which are bound to the chiral atom, confirming the structures of synthesized compounds **3a**, **3b**, and **3c**. These protons in the ^{1}H NMR spectra appear as a doublet of doublets with vicinal and geminal constants that characterize the individual state of each compound. The diffuse reflectance solid-phase IR spectra make it possible to interpret the changes in the positions of the characteristic frequencies as a consequence of intermolecular interactions [4]. Of the synthesized compounds, compound **3c** has an analogous property, and in its spectrum the O-H frequency of the phenolic hydroxyl shifts to the long-wavelength region of the spectrum (3189 cm^{-1}). In a solution of compound **3c**, the O-H frequency of the hydroxyl appears at 3435 cm^{-1}. It can be assumed that in a molecule of compound **3c** the hydrogen atom interacts with the oxygen atom of the functional group of the *para*-substituent and this interaction is accompanied by the enolation of the aromatic bonding system, which agrees with the IR spectra data of cyclohexadienones [5, 6]. The IR spectrum of salt **3c** contains intense bands at 1667, 1632, and 1596 cm^{-1}, whereas the band at 1550 cm^{-1} characteristic of the aromatic structure is absent. A similar effect related to the change in the position of the characteristic frequencies is manifested in the IR spectra of compound **2a** and corresponding potassium salt **4**. The IR spectra of these compounds contain no frequencies of the carboxyl group at 1700 cm^{-1} but have a broad band at 1550 - 1620 cm^{-1}. These data can be interpreted as a consequence of the interaction of the sodium (potassium) cation with π-electrons of the atoms of the six-membered ring and oxygen atoms of the functional groups. The data of quantum chemical calculation of structures **2a**, **3a**, and **4** in the PM6 approximation (Mopac 2007 program) [7] confirm the participation of the metal cation in the coordination bond with the aromatic bonding system and oxygen atoms of the *para*-substituent (Figure 4).

Thus, the above results show that the functional group of the *para*-substituent of sterically hindered phenol exerts a specific effect on the properties of the aromatic bonding system, which should manifest itself in the antioxidation properties of the synthesized compounds. The properties of the antioxidants depend on the reaction constants of the peroxide radicals with phenol and the backward reaction of the phenoxyl radical with hydroperoxide, which made it possible to use the calculation technologies for the prediction of the efficiency and choice of optimal structures for the synthesis [8]. For this purpose, the energies of homolysis of the O-H bond of phenolic hydroxyl (D_{OH}) were calculated: $D_{OH} = \Delta H(AlkArO) + \Delta H(H) - \Delta H(AlkArOH)$, where $\Delta H(AlkArO)$ is the energy of formation of the phenoxyl radical, and $\Delta H(AlkArOH)$ is the energy of formation of phenol (compounds 2a, 2d, 3b, 3c, and 4). The results are listed in Table 1.

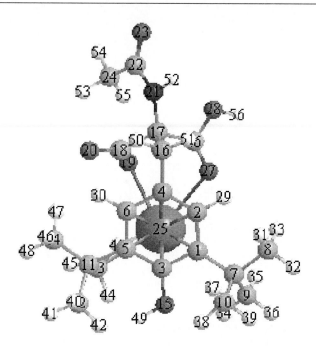

Figure 4. Structure of compound 2a according to the data of the PM6 calculation.

**Table 1. Energies of homolysis of the O—H bond of phenolic hydroxyl
in 2d, 3a—c, and 4 and cumene hydroperoxide**

Compound	$D_{(OH)}$/kcal mol^{-1}	$D_{(OH)}$/kJ mol^{-1}
3b	80.0	334.7
2d	76.6	320.5
3c	76.6	320.5
3a	71.7	300.0
4	74.6	312.1
Cumene hydroperoxide	75.2	314.6

Cumene-soluble compounds **2d**, **3b**, and **3c** were used in the estimation of the influence of the intramolecular interaction on the antioxidation parameters: the period of oxidation inhibition involving the antioxidant (τ) and the chain termination factor (f) at the constant initiation rate $W_i = 1.5 \cdot 10^{-8}$ m L^{-1} s^{-1}. These parameters were related by the expression: $f = \tau W_i$[InH]$^{-1}$, where InH are compounds **3b**, **3c**, or **2d**. The results of determination of the induction period of cumene oxidation in the presence of InH are shown in Figure 5.

It follows from these results that the antioxidation properties (by the f value) increase in the series of compounds 2d > 3b > 3c, whereas similar values of the f factor for compounds 2d and 3c should be expected from the results of calculation of the energy of homolysis of the O-H bond.

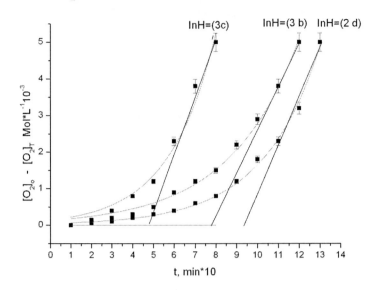

Figure 5. Kinetic curves of oxygen absorption in the initiated oxidation of cumene at 50 □C in the presence of compounds 3c (1), 3b (2), and 2d (3) and azodiisobutyronitrile (AIBN) (4); Wi = 1.5•10– mol L–1 s–1, C0.mol L–1 = 2.4•10–5 (1); 3.1•10–5 (2); 3.3•10–5 (3); f = 1.71 (1); 1.98 (2); 2.42 (3).

In our opinion, this contradiction is due to taumerism of compound 3c between the aromatic and spirocyclic structures (Figure 6), resulting in a decrease in the concentration of the aromatic structure under the experimental conditions. It follows from the energies of formation of aromatic structure 3c (A) (–883.544 kJ) and spirocyclic structure 3c (B) (–841.748 kJ) that the difference is 41.796 kJ mol^{-1} (9.98 kcal mol^{-1}). In the nonpolar solvent (cumene) structure 3c (B) is more stable, because the dipole moment (D) of structure 3c (A) is 3.47 D, which is lower than D 3c (A) = 6.13 D.

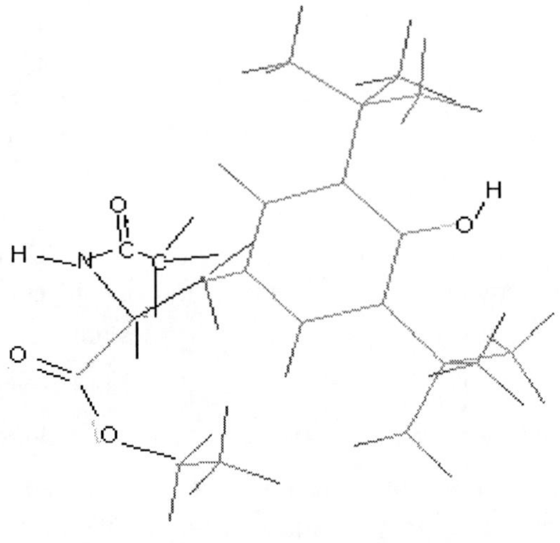

структура 3c A

Figure 6. (Continued on next page)

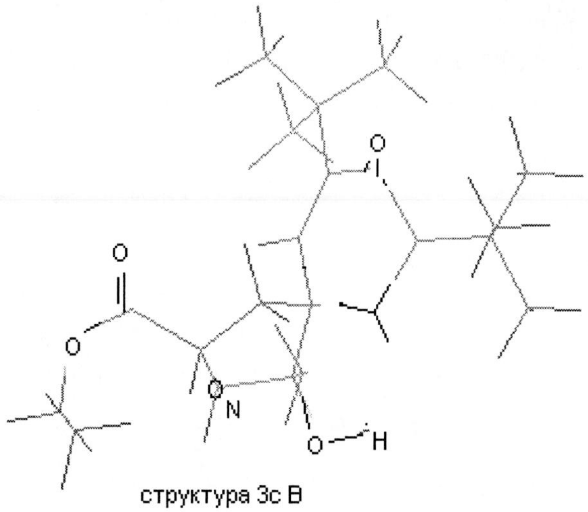

структура 3с В

Figure 6. Structures of tautomers 3c (A and B) according to the data of quantum chemical calculations in the PM6 approximation.

The data in Table 1 show that compounds **2d**, **3b**, and **3c** are insufficiently efficient antioxidants, whereas compounds **3a** and **4** can be interest as water-soluble and efficient antioxidants.

EXPERIMENTAL

The parameters of the structures of the synthesized compounds were calculated using the Mopac-2007, Version 8.288W program in the PM6 approximation. ^{1}H NMR spectra were recorded on a Bruker WM-400 instrument (400 MHz) relative to the signal of residual protons of the deuterated solvent (acetone-d_6 or DMSO-d_6). IR diffuse reflectance spectra were recorded on a Perkin-Elmer 1725-X spectrometer for crystals. The antioxidation parameters of the synthesized compounds were determined by a known method [9] under the conditions of inhibited oxidation of cumene with oxygen at 50 °C in the presence of the oxidation initiator (azodiisobutyronitrile).

Diethyl N-acetylamino(3,5-di-tert-butyl-4-hydroxybenzyl)malonate (1) was synthesized by a known procedure [6] m.p. 130 °C. ^{1}H NMR (acetone-d_6), δ: 1.25 (t, 6 H, C$_2$H$_3$, J = 7.1 Hz); 1.40 (s, 18 H, Bu^t); 2.07 (s, 3 H, COCH$_3$); 2.89 (s, 1 H, CH$_2$); 4.20—4.23 (m, 4 H, CH$_2$CH$_3$); 6.01 (s, 1 H, OH); 6.82 (s, 2 H, Ar); 7.29 (br.s, 1 H, NH).

Alkaline hydrolysis of compound 1. Compound **1** (4.36 g, 0.01 mol) was added to a solution of NaOH (0.84 g, 0.06 mol) in dioxane (25 mL) and water (3 mL) at 100 °C in an argon flow, and samples for analysis were taken at an interval of 30 min for 3 h under the temperature-controlled conditions. The composition of the reaction mixtures was monitored (after neutralization to pH 4) using the comparison of the ^{1}H NMR spectra in an acetone-d_6 solution by the signals from *meta*-protons of the aromatic ring of the initial ester **1** (δ_{f} 6.820) and reaction products: δ_{f} 7.05 (**3b**), δ_{f} 6.95 (**3c**), and δ_{f} 6.92 (**2d**).

Synthesis of monosodium salt of N-acetylamino(3,5-di-tert-butyl-4-hydroxy-benzyl)malonic acid (2a) and 2-N-acetylamino-3-(3,5-di-tert-butyl-4-hydroxyphenyl)-propanoic acid (3b). A mixture of compound **1** (43.6 g, 0.1 mol), NaOH (4 g, 0.1 mol), water (20 mL), and dioxane (200 mL) was heated for 2 h at 85—90 °C in an argon flow. Then the reaction mixture was cooled down, and the precipitate formed was separated and recrystallized from aqueous ethanol (8:2). Crystalline hydrate of salt **2a** was obtained in a yield of 23.4 g (62%), m.p. 258—260 °C (with decomp.). ^{1}H NMR (DMSO-d_6), δ: 1.26 (s, 18 H, Bu^{t}); 1.79 (s, 3 H, COCH$_3$); 3.37 (s, 2 H, CH$_2$); 3.40—3.48 (br.s, HOH); 6.42—6.46 (br.s, 1 H, OH); 6.79 (s, 2 H, Ar); 7.20—7.27 (br.s, 1 H, NH). IR, v/cm^{-1}: 3643 (OH); 3550—3100 br. (HOH); 3321 (NHCOCH$_3$); 2957 (CH); 1550—1615 br. (COOH, COO$^-$, C=C); 1507; 1433; 1417; 1287; 1234; 1215; 1156; 1123. Found (%): C, 56.96; H, 7.37; N, 3.52; Na, 5.55. C$_{20}$H$_{28}$NO$_6$Na•H$_2$O. Calculated (%): C, 57.27; H, 7.20; N, 3.34; Na, 5.48.

The mother liquor was added by 10% HCl to pH 4, the solvent was evaporated, and the residue was crystallized from acetone. Compound *3b* was obtained in a yield of 5.37 g (17%), m.p. 205—206 °C (from EtOH). According to the literature data m.p. 203 °C. ^{1}H NMR (acetone-d_6), δ: 1.42 (s, 18 H, Bu^{t}); 1.96 (s, 3 H, COCH$_3$); 2.94 (dd, 1 H$_{\text{à}}$, J = 7.6 Hz); 3.11 (dd, 1 H$_{\text{b}}$, J = 5.0 Hz); 4.63—4.69 (m, 1H$_{\tilde{\text{n}}}$); 5 .90 (s, 1 H, OH); 7.05 (s, 2 H, Ar); 7.45 (d, 1 H, NH, J = 5.0 Hz). IR, v/cm^{-1}: 3639 (OH); 3332 (NHCOCH$_3$); 2954, 2913, 2872 (CH); 1715 (COOH); 1624 (HNCO); 1550 (C=C), 1433, 1269, 1218, 1188, 1157, 1120. Found (%): C, 67.84; H, 8.85; N, 4.16. C$_{19}$H$_{29}$NO$_4$. Calculated (%): C, 68.03; H, 8.72; N, 4.18.

Monoethyl N-acetylamino(3,5-di-tert-butyl-4-hydroxybenzyl)malonate (2c). A mixture of compound **1** (43.6 g, 0.1 mol) and AcONa (8.2 g, 0.1 mol) in water (30 mL) and propan-2-ol (200 mL) was stored for 3 days at room temperature, after which 10% HCl (35 mL) was added. Compound **2c** was obtained in a yield of 26.1 g (64%), m.p. 175—176 ° (from toluene). ^{1}H NMR (DMSO-d_6), δ: 1.24 (t, 3 Í, CH$_2$CH$_3$, J = 7.1 Hz); 1.34 (s, 18 H, Bu^{t}); 1.89 (s, 3 H, COCH$_3$); 3.22 (s, 2 H, CH$_2$); 4.13 (2 H, CH$_3$CH$_2$, J =7.1 Hz); 6.34 (s, 1 H, OH); 6.76 (s, 2 H, Ar); 7.39 (s, 1 H, NH). IR, v/cm^{-1}: 3633 (OH); 3345 (NHCOOCH$_3$); 2954 (CH); 1749 (COOC$_2$H$_5$); 1714 (COOH); 1619 (HNCOCH$_3$); 1533 (C=C); 1435, 1235, 1212, 1149. Found (%): C, 64.89; H, 8.33; N, 3.31. C$_{22}$H$_{33}$NO$_6$. Calculated (%): C, 64.85; H, 8.16; N, 3.44.

Ethyl 2-N-acetylamino(3,5-di-tert-butyl-4-hydroxyphenyl)propanoate (3c) is formed on heating monoethyl ester **2c** to temperatures higher than 176 °C due to decarboxylation. M.p. 135—136 °C (from toluene). ^{1}H NMR (acetone-d_6), δ: 1.68 (t, 3 H, CH$_3$CH$_2$, J = 7.1 Hz); 1.42 (s, 18 H, Bu^{t}); 1.91 (s, 3 H, COCH$_3$); 2.02 (dd, 1 H$_{\text{à}}$, J = 6.2 Hz); 3.11 (dd, 1 H$_{\text{b}}$, J = 6.3 Hz); 4.09 (2 Í, CH$_3$CH$_2$, J = 7.1 Hz); 4.60—4.64 (m, 1H$_{\tilde{\text{n}}}$); 5.98 (s, 1 H, OH); 6.95 (s, 2 H,

Ar); 7.34 (d, 1 H, NH, J = 7.15 Hz). IR, v/cm^{-1}: 3354 (N<u>H</u>COCH$_3$); 3189 br. (OH); 2948 (CH); 1732 (<u>C</u>OOC$_2$H$_5$); 1647 (HN<u>C</u>O); 1548 (C=C); 1435; 1253; 1211; 1039. Found (%): C, 69.59; H, 9.13; N, 3.94; C$_{21}$H$_{33}$NO$_4$. Calculated (%): C, 69.40; H, 9.15; N, 3.85.

N-Acetylamino(3,5-di-tert-butyl-4-hydroxybenzyl)malonate (2d). A solution of compound **2a** (4.2 g, 0.01 mol) in water (50 mL) was added by 10% HCl (5 mL). The precipitate formed was separated, dried at 25—30 °C, and crystallized from a toluene—EtOH (9 : 1, vol.) mixture. The yield was 92—96%, m.p. 198—200 °C (from toluene). According to the literature data [1] m.p. 148 °C. ^{1}H NMR (DMSO-d$_6$), δ: 1.34 (s, 18 H, Bu^t); 1.87 (s, 3 H, COCH$_3$); 3.22 (s, 2 H, CH$_2$); 6.74 (s, 2 H, Ar); 7.69 (s, 1 H, NH). IR, v/cm^{-1}: 3637 (OH); 3338 (N<u>H</u>COCH$_3$); 2954, 2913, 2872 (CH), 1716 (<u>C</u>OOH); 1623 (HN<u>C</u>OCH$_3$); 1536 (C=C), 1435, 1214, 1155, 1121. Found (%): C, 63.24; H, 7.95; N, 3.86. C$_{20}$H$_{29}$NO$_6$. Calculated (%): C, 63.31; H, 7.70; N, 3.69.

N-Acetylamino(3,5-di-tert-butyl-4-hydroxybenzyl)malonic acid diammonium salt (2e). Ammonia (2.2 mL, 0.1 mol) was added to a solution of acid *2d* (3.37 g, 0.01 mol) in EtOH (15 mL). The reaction mixture was stirred for 30 min, and ammonia and solvent excess were distilled off. Water (10 mL) was added to the residue, and the mixture was heated and filtered. The filtrate was cooled to 5—6 °C and stored until precipitation, and the precipitate was separated by filtration. Compound **2e** was obtained in a yield of 3.1 g (83%), m.p. 177—178 °C(from EtOH). ^{1}H NMR (DMSO-d$_6$), δ: 1.29 (s, 18 H, Bu^t); 1.75 (s, 3 H, COCH$_3$); 2.52 (s, 2 H, CH$_2$); 3.85—4.01 (br.s, 8 H, NH$_4$); 6.80 (s, 2 H, Ar); 7.04—7.06 (br.s, 1 H, NH). Found (%): C, 57.95; H, 8.71; N, 10.26. C$_{20}$H$_{35}$N$_3$O$_6$. Calculated (%): C, 58.01; H, 8.53; N, 10.16.

N-Acetylamino(3,5-di-tert-butyl-4-hydroxybenzyl)malonic acid monopotassium salt (**4**) was synthesized similarly to *2a* in 56% yield, m.p. 250—252 °C (with decomp.). ^{1}H NMR (DMSO-d$_6$), δ: 1.27 (s, 18 H, Bu^t); 1.80 (s, 3 H, COCH$_3$); 2.48 (s, 2 H, CH$_2$); 3.41—3.46 (br.s, 3 H,HOH); 6.42—6.47 (br.s, 1 H, OH); 6.79 (s, 2 H, Ar); 7.22—7.26 (br.s, 1 H, NH). IR, v/cm^{-1}: 3644 (OH); 3550—3100 br. (HOH); 3323 (N<u>H</u>COCH$_3$); 1550—1620 br. (HN<u>C</u>O, <u>C</u>OO$^-$, C=C). Found (%): C, 54.04; H, 7.03; N, 3.15; Ê, 8.60. C$_{20}$H$_{22}$KNO$_6$•1.5H$_2$O. Calculated (%): C, 54.32; H, 7.21; N, 3.06; K. 8.75.

2-N-Acetylamino(3,5-di-tert-butyl-4-hydroxyphenyl)propanoic acid sodium salt sodium salt (3a). A. a solution of compound *3b* (3.16 g, 0.01 mol) in water (10 mL) was added by NaOH (0.4 g, 0.01 mol). The mixture was heated until the precipitate dissolved, and then the solvent was distilled *in vacuo*. The residue was added by EtOH (10 mL), NaCl was separated by filtration, the mother liquor was evaporated, and the crystals of compound *3a* were washed with EtOH and dried in a desiccator over P$_2$O$_5$. Compound *3a* decomposes on heating above 250 °C. ^{1}H NMR (DMSO-d$_6$), δ: 1.32 (s, 18 H, Bu^t); 1.75 (s, 3 H, COCH$_3$); 2.69 (dd, 1 H$_à$, J = 7.6 Hz); 2.94 (dd, 1 H$_b$, J = 5.0 Hz); 4.01—4.06 (m, 1 H$_ñ$); 3.52—3.57 (br.s, H, H$_2$O); 6.96—6.98 (br.s, 1 H, OH); 6.87 (s, 2H, Ar); 7.43—7.45 (br.s, 1 H, NH). IR, v/cm^{-1}: 3646 (OH); 3285 br. (HOH); 3103 (N<u>H</u>COCH$_3$); 2956, 2914, 2874 (CH); 1667 (C=O); 1632

(HNCOCH$_3$); 1596 (C=C); 1435; 1400; 1316; 1400; 1316; 1234; 1158; 1121. Found (%): C, 55.64; H, 8.15; N, 3.46; Na, 5.64. C$_{19}$H$_{28}$NO$_4$Na•3H$_2$O. Calculated (%): C, 55.45; H, 8.34; N, 3.40; Na, 5.59.

B. A solution of compound **2a** (4.2 g, 0.01 mol) in water (20 mL) was heated to boiling and stored for 35—40 min, then the solvent was evaporated, and the residue was crystallized from an EtOH - water (1:1) mixture. The yield of compound **3a** was 3.78 g (92%). The ^{1}H NMR spectrum of the sample coincides with that of compound **3a** obtained by method *A*.

CONCLUSION

The efficiency of process alkaline hydrolysis of diethyl *N*-acetylamino(3,5-di-*tert*-butyl-4-hydroxybenzyl)malonate depends on the temperature and ratio of the reactants, that is accompanied by decarboxylation.

REFERENCES

[1]   H. J. Teuder, H. Rrause, V. Berariu, «Syntesis of 3,5-di-tert-butyltyrosin and (3,5-di-tert-buthy-4-hydrophenyl)glycin», Lieb. Ann., 1978, 757.

[2]   C. W. Gear, «Numerical Initial Value Problems in Ordinary Differential Equations», Prentice Hall, New York, 1971, 158.

[3]   N. B. Colthup, L. H. Daly, S. E. Wiberley, «Introduction to Infrared and Raman Spectroscopy», Àcademic Press, New York—London, 1990, 319.

[4]   S.M.Rentzel-Edens, V. A. Russell, Lian Yu, «Molecular basis for the stability realationships between homochiral and racemic crystals of tazofelone: a spectroscopic, critallographic and thermodinamic investigation», J. Chem. Soc., Perkin Trans. 2, 2000, 913.

[5]   A. A. Volod´kin, R. D. Malysheva, V. V. Ershov, «Oxidation of 4-substution 2,6-di-tert-buthylphenols by oxygen into alkaline solution», Izv. Akad. Nauk SSSR, Ser. Khim., 1982, 1594 [Bull. Acad. Sci. USSR, Div. Chem. Sci. (Engl. Transl.), 182, 1633].

[6]   R. M. Guerra, I. Duran, P. Ortiz, «Chemical Reaction: Quantitative Level of Liquid and Solid Phase», Nova Science Publishers, Inc., New York, 1992, pp. 59—74.

[7]   J. J. P. Stewart , «Computational», J. Mol. Mod., 2007, **13**, 1173.

[8]   D. V. Berdyshev, V. P. Glazunov, V. L. Novikov, «Antioxydant propeties of 2,3,6,8-pentahydroxy-7-ethyl-1,4-naphtohinon and theorethycal investigation by quanto-chemistry»,Izv. Akad. Nauk, Ser. Khim., 2007, 400 [Russ. Chem. Bull., Int. Ed., 2007, **56**, 413].

[9]   N. M. Emanuel, E. T. Denisov, Z. K. Maizus, «Tsepnye reaktsii okisleniya uglevodorodov v zhidkoi faze [Chain Reactions of Hydrocarbon Oxidation in the Liquid Phase]», Nauka, Moscow, 1965 (in Russian).

In: Polymer Yearbook – 2011.
Editors: G. Zaikov, C. Sirghie et al. pp. 131-136

ISBN 978-1-61209-645-2
© 2011 Nova Science Publishers, Inc.

Chapter 12

MECHANISM OF CATALYTIC ALKYLATION OF 2,6-DI-TERT-BUTYLPHENOL BY METHYL ACRYLATE

A. A. Volod'kin and G. E. Zaikov

N.M. Emanuel Institute of Biochemical Physics Russian Academy of Sciences[1]
Moscow 119334, Russia

ABSTRACT

The determining factor of the reaction of 2,6-di-*tert*-butylphenol with alkaline metal hydroxides is temperature, depending on which two types of potassium or sodium 2,6-di-*tert*-butyl phenoxides are formed with different catalytic activity in the alkylation of 2,6-di-*tert*-butylphenol with methyl acrylate. More active forms of 2,6-$\text{Bu}^t_2\text{C}_6\text{H}_3\text{OK}$ or 2,6-$\text{Bu}^t_2\text{C}_6\text{H}_3\text{ONa}$ are synthesized at temperatures higher than 160 °C and represent predominantly monomers of 2,6-di-*tert*-butyl phenoxides producing dimers on cooling.

Keywords: phenols, phenoxides, 2,6-di-tert-butylphenol, methyl acrylate, Michael reaction, kinetics.

INTRODUCTION

Phenoxides based on sterically hindered phenols and alkali metals are used in the synthesis and industrial production of a number of compounds applied in the production of antioxidants [1].

The shielding of the reaction center by the tert-butyl groups gives rise to the specific properties inherent in sterically hindered phenoxides, which are related to negative charge

---

[1] 4 Kosygin str., Moscow 119334, Russia,chembio@sky.chph.ras.ru.

delocalization in the six-membered ring and the formation of a coordination bond of the metal cation with the oxygen atoms and carbon atoms of the six-membered ring. These feature are responsible for the ability of potassium (1) and sodium (2) 2,6-di-tert-butylphtnoxides to form supramolecular structures as contact or solvent-separated ion pairs and dimmers [2]. Similar structures based on alkali metal cations and radical anions have been studies previously by EPR [3]. The above data suggest that a nonsolvated cation may by formed in a matrix comprising 2.6-di-tert.-butylphenol (3), methylacrylate (4), and catalytic amounts of monomer 1 or 2. The course of alkylation of phenol 3 with methyl acrylate in the presence of catalytic amount of monomer 1 or 2 differs from that in the presence of the corresponding phenoxide dimmers (associates).

EXPERIMENTAL

PMR ^{1}H spectra were recorded on a Bruker WM-400 instrument (400 MHz) in DMSO-d_6. Molecular weights were determined by the thermoelectric method using a cell described previously and calculated by the formula Equation: $M = (k_g M_s / 100(t_1 - t_2))$, where g is a weighed sample (g) per 100 g of solvent (2—3%), k is the coefficient of the instrument and thermistors, M_s is the molecular weight of a standard sample, and $t_1 - t_2$ is the value equivalent to the temperature change in the cell filled with a sample solution by the potentiometer scale. The temperature in the thermostat was 82 °C. Kinetic data were obtained by liquid chromatography of reaction mixtures (Bruker LC-31 chromatograph, IBM Cyano column, hexane − isopropyl alcohol ethyl acetate (8:1:1) mixture as eluent, rate 0.4 mL min^{-1}).

A. Sodium 2,6-di-tert-butylphenoxide (2) a Flask was filled with 3 (20.6 g, 0.1 mol) and heated in an argon flow to 190 °C, and granulated NaOH (1.4 g) was added. After 10 min crystal that formed were separated by high-temperature filtration under argon and washed with n-octane heated to 110—115 °C until no $2,6\text{-But}_2\text{C}_6\text{H}_3\text{OH}$ was found in the mother liquor.

Residues of the solvent were separated in vacuo at 120—130 °C to obtain 2 (5.3 g, 93.5%), which was placed under argon into a flask heated to 120—130 °C. NMR ^{1}H, δ: 1.30 (s, 18 H); 5.58 (t, 1 H, J = 6 Hz); 6.58 (d, 2 H, J = 6 Hz). PMR ^{1}H spectrum of 2,6-di-tert-butylphenol (for comparison) , δ: 1.37 (s, 18 H); 3.48 (s, 1 H); 6.74 (t, 1 H, J = 8 Hz); 7.06 (d, 2 H, J = 8 Hz). Found: M $_{min}$ = 234. Calculated for $2,6\text{-But}_2\text{C}_6\text{H}_3\text{ONa}$: M = 228.96.

B. A flask was filled with 3 (20.6 g, 0.1 mol) and heated in an argon flow to 190 °C, and granulated NaOH (1.4 g) was added. After 10 min the reaction mixture as a suspension was cooled to ~20 °C, and heptane (50 mL) was added. The precipitate was filtered off and washed with heptane until no 3 was found in the washing solutions to obtain 4.9 g (~88%) of dimmer 2; M $_{max}$ = 458.

Methyl 3-(-3,5-di-tert-butyl- 4-hydroxyphenyl) propionate (5).

A.   A Methyl acrylate (2.5 mL, 0.03 mol) was added in an argon flow to a mixture of 2 *(method A)* and 3 (4.88 g, 0.01 mol) at 115 °C. After 30min the reaction mixture was cooled to ~20 °C, 10% HCl was added to neutral pH, and the product was extracted with hexane. The compound **5** was obtained in a yield of 5.66 g (96%), m.p. 66 °C.

B. Granulated KOH (0.11 g, 0.002 mol) was added in an argon flow **3** (20.6 g, 0.1 mol) heated to 190 °C. After 10 min the reaction mixture was cooled to 130 °C, and methyl acrylate (11 g, 0.13 mol was added. After the reactants were mixed, the temperature of the reaction mixture decreased to 110 °C. After 25 min the content of the product in the reaction mixture was 98 mol.%.

C. Methyl acrylate (11 g, 0.13 mol) was added in an argon flow at 110 °C v to a mixture of **3** (20.6 g, 0.1 mol) and **2** (0.488 g, 0.002 mol) synthesized by method B. After 25 min the content of the product in the reaction mixture became 15 mol.%, while after 3 h it reached 72%. Similarly, when **1** (0.46 g, 0.002 mol) synthesized by method **B** was used, the product yield after 3 h became 86%.

INFLUENCE OF SOLVENT ADDITIVES ON THE CATALYTIC ALKYLATION OF 3 WITH METHYL ACRYLATE IN PRESENCE OF 2

Granulated NaOH (0.12 g, 0.002 mol) was added to **3** (20.6 g, 0.01 mol) at 190 °C an argon flow. After 10 min , the reaction mixture was cooled to 135 °C, and methyl acrylate (11.2 g, 0,13 mol) and DMCO (0.2 mL) were added. The reactions with additives HMPA, DMF, MeCN and Diglime were carried out similarly.

RESULTS AND DISCUSSION

In this work, we found that, with a nearly equimolar mixture or 3 and 4 , the catalytic alkylation in the presence of 1.5-3 mol % of monomer 1 or 2 proceeds at higher rates than those known previously (in the presence of dimmer 1 or 2) [2 - 5] and gives the alkylation product methyl 3-(3,5-di-tert-butyl-4-hydroxyphenyl)propionate (5), in up to 98% yield. The formation of the monomer species was supported by thermoelectric measurements of the molecular weights of phenoxides 1 and 2 at 115 °C. The kinetic studies for this reaction in the range 105-130 °C showed that the pattern of the kinetic curve does not depend on the reaction temperature (Figure1), which may imply a tunneling effect in the catalytic alkylation mechanism.

The monomer of phenoxide 1 is formed upon the reaction of phenol 3 with potassium hydroxide at 180-200 °C. On cooling to 20 °C, the monomer is converted into dimer 1. In the experiment, a mixture of 1 and 3 was cooled to 135-120 °C and compound 5 was simultaneously added. This gave a reaction mass in which phenoxide 1 or 2 was mainly uniformly distributed throughout the mixture of phenol 3 and acrylate 4 and the reaction between these compounds proceeded under steady-state conditions in a thermostat without stirring the reactants.

By sampling the reaction mixture during the process and analyzing the samples by liquid chromatography, the contents of 3,5 and dimethyl − (3,5-di-tert-butyl-4-hydroxy-benzyl)glutarate (6) in the reaction mixture were the presence of polar or proton-containing solvent additives , and other factors that can affect the catalyst properties. When the initial 3 to 4 ratio is nearly equimolar and 1 or 2 is present in catalytic amounts , the reactions take place between 1 (2) and 3 and between 1 (2) and 4 to give the alcylation products 5 and 6 and methyl acrylate oligomers. When the ratio 4:3>3, the yield of alkylation product 5 decreases to 10-15%. Presumably , with excess methyl acrylate , the reaction matrix properties change and the nonassociated cations enter the bulk and react with methyl acrylate to give oligomers. In a binary reactions mass consisting of phenoxide 1 in a methyl acrylate solution, oligomerization predominates.

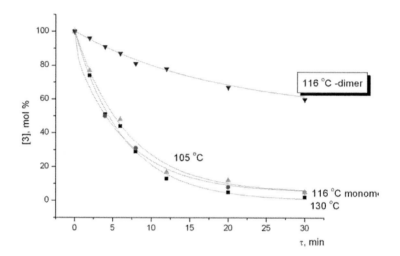

Figure1. Temperature dependence of the kinetics of the reaction of phenol **3** with methyl acrylate in presence **1**. $[3]_o$= 3.29, $[4]_o$ =3.75 $[1]_o$ = 0.108 mol.L^{-1}; the dots correspond to the experiment.

Table 1. Rate constant for the elementary steps of the catalytic alkylation of 2,6-di-tert-butylphenol (3) by methyl acrylate (4) in the presence of sodium 2,6-di-tert-butylphenoxid (2)

Reaction	№	k
PhONa --->PhO$^-$ + Na$^+$	(1)	$4,5.10^{-2}$ c^{-1}
PhO$^-$ + Na$^+$ ---> PhONa	(2)	1.10^{-2} L.mol^{-1}.c^{-1}
PhO$^-$ + MA ---> ArO$^-$	(3)	$1.8.10^{-3}$ L.mol^{-1}.c^{-1}
Na$^+$ + PhOH ---> PhONa +H+	(4)	$1,5.10^{-1}$ L.mol^{-1}.c^{-1}
PhONa + H$^+$ ---> Na$^+$ + PhOH	(5)	5.10^{-2} L.mol^{-1}.c^{-1}
H$^+$ + ArO$^-$ --->ArOH	(6)	7.10^{-1} L.mol^{-1}.c^{-1}
Na$^+$ +MA + MA ---> Na$^+$2MA	(7)	3.10^{-4} L^2.mol^2.c^{-1}
PhO$^-$ + MA + MA ---> Ar'O$^-$	(8)	1.10^{-5} L^2.mol^2.c^{-1}
Ar'O$^-$ + H$^+$ ---> Ar'OH	(9)	7.10^{-1} L.mol^{-1}.c^{-1}
Na$^+$ + MeCN ---> Na$^+$MeCN	(10a)	5.10^{-3} L.mol^{-1}.c^{-1}
Na$^+$ + HMPA ---> Na$^+$HMPA	(10b)	$2.2.10^{-2}$ L.mol^{-1}.c^{-1}
Na$^+$ + DMF ---> Na$^+$DMF	(10c)	$4.2.10^{-2}$ L.mol^{-1}.c^{-1}
Na$^+$ + DMSO ---> Na$^+$DMSO	(10d)	$6.4.10^{-2}$ L.mol^{-1}.c^{-1}
Na$^+$ +Diglyme ---> Na$^+$Diglyme	(10e)	$6.1.10^{-1}$ L.mol^{-1}.c^{-1}

MeCN is acetonitrile; HMPA is hexcamethylphosphoramide; DMF is dimethylformamide; DMSO is dimethyl sulfoxide; Diglyime is dimethoxyethane.

This gives an oligomer with a molecular mass of > 10000 containing a phenolic fragment. On the basis of the foregoing , a scheme of the catalytic alkylation of 2,6-di-tert-butylphenol with methyl acrylate was proposed. In terms of this scheme, mathematical modeling was carried out using the experimental kinetic data for various initial concentrations of 3,4, and 2.

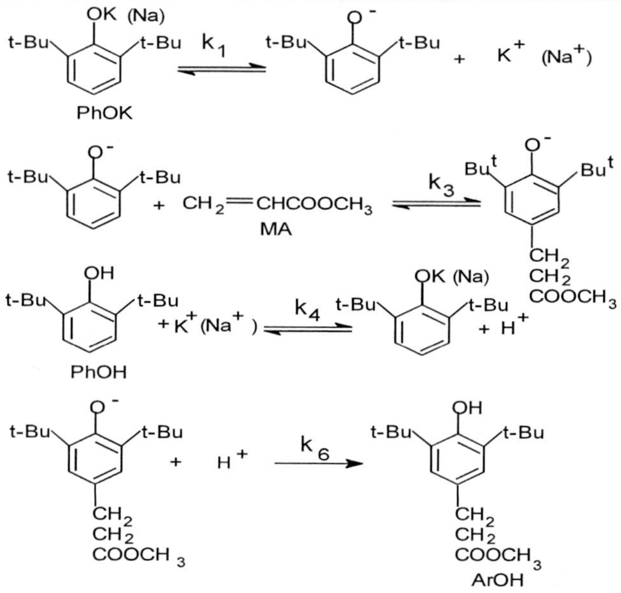

Scheme 1.

The modeling was carried out using a computer program for reaction kinetics based on the solution of a set of differential equations. The kinetic scheme consists of three blocks (table). The first block {reactions (1)-(6)} includes the steps of formation of the alkylation product **5** (the numbers of the constants correspond to the numbers of reactions in the table).

The second block contains two reactions, resulting in product **6** and methyl acrylate oligomer. The third block is related to the inhibition caused by the reaction of the metal cation with the solvent additive (the formation of an associated cation, a crown ether-type complex, or dimmer associated). The stimulating influence of polar solvents in bimolecular addition is known [6].

However, in this case ,polar solvents have an adverse effect on the catalytic alkylation which is manifestd as a decrease in the consumption rate of phenol **3** and a decrease in the yield of **5** (Figure2).

This fact is at variance with the classical mechanism; however it is consistent with the ion chain mechanism because the the formation of a nonassociated metal cation implied regeneration of the catalyst **(1 or 2)**. In the series of solvents studies, dimethoxyethane is most efficient additive, having an inhibiting effect due to formation of a crown ether-type complex (cation trap).

CONCLUSION

Thus, experimental data are consistent with the possibility of the ion chain mechanism in the catalytic alkylation of 2,6-di-tert-butylphenol with methyl acrylate in presence of potassium and sodium 2,6-di-tert-butylphenoxide monomers.

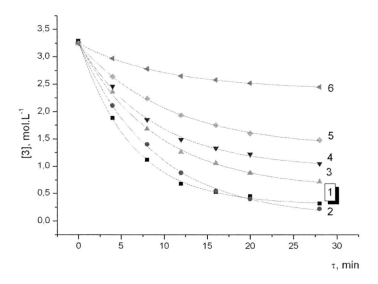

Figure 2. Kinetics of reaction of phenol **3** with acrylate (**4**) in presence of phenoxide **2** and solvent additives (solvent, mol..L^{-1}) 110 $^\circ$C; (1) no solvent; 2 (MeCN), 0.24; (3) HMPA, 0.22; (4) DMF; 0.21; (5) DMSO, 0.23; (6) diglyme, 0.19. $[3]_o$ = 3.26-3.28; $[4]_o$ = 3.5-3.6; $[2]_o$ = 0.1 mol.L^{-1}; the curves show the calculation data and the dots correspond to the experiment.

REFERENCES

[1]   A.A.Volod'kin, V.I.Paramonov, F.M.Egidis L.K.Popov , «The Influence of a catalysts on rate of alcylation 2,6-di-tert-butylphenol by methyl acrylate» Chemical Industry 1988, №12, pp.7-11.

[2]   A.A.Volod'kin, A.S.Zaitsev, V.L.Rubailo, V.A.Beleakov, G.E.Zaikov, «The Reaction mexanism of 2,6-di-tert-butylphenol alkylation by methyl acrylate in presence of 2,6-di-tert-butyl phenolate of potassium and alkali», Polymer Degradation and Stability, 1989, <u>26</u>, pp.69-100.

[3]   A.I.Prokof'ev,S.P.Solodovnikov, A.A.Volod'kin, V.V.Erschov, «The investigation of pare- ion on 4,4-dimethyl-2,6-di-tert-butylcycloxechsadien-2,5-ona – sodium» Izv.Akad.NaukSSSR, Ser.Chim., 1972, pp.640-643.

[4]   A.A.Volod'kin, G.E.Zaikov, «Kinetic and mexanism of alkylation 2,6-dialkylphenols by unsuturating compounds» Ross.Khim.Zh. ,2000, vol.**44**, № 26, pp.81-89.

[5]   A.A.Volod'kin, G.E. Zaikov, «Formation and properties of a catalyst based on sodium and potassium hydroxides in the reaction of 2,6-di-tert-butylphenol with methyl acrylate», Russian chemical Bulletin, International Edition, 2002 , <u>51</u>, 2189-2195.

[6]   G.Becker, Einfuhrung in die Elektronentheorie organischemischer Reaktionen, Berlin, Deutscher Verlag der Wissenschaften, 1974,

In: Polymer Yearbook – 2011.
Editors: G. Zaikov, C. Sirghie et al. pp. 137-145

ISBN 978-1-61209-645-2
© 2011 Nova Science Publishers, Inc.

Chapter 13

THE THEORETICAL STRUCTURAL MODEL OF NANOCOMPOSITES POLYMER/ORGANOCLAY REINFORCEMENT

G. V. Kozlov, [1)] *B. Zh. Dzhangurazov,* [2)]
G. E. Zaikov [1)] *and A. K. Mikitaev*
[1)]Kabardino-Balkarian State University,[1]
Nal'chik-360004, Russian Federation
[2)]N.M. Emanuel Institute of Biochemical Physics
of Russian Academy of Sciences, [2]
Moscow-119334, Russian Federation

ABSTRACT

The simple theoretical model of nanocomposites polymer/organoclay reinforcement was proposed. Unlike the existing micromechanical models the offered treatment takes into account real nanofiller – polymeric matrix adhesion level. It has been shown that interfacial regions are reinforcing elements togther with nanofiller. These two structural components form nanoclay "effective particle".

Keywords: Nanocomposite, organoclay, interfacial regions, adhesion, reinforcement degree.

INTRODUCTION

At present nanocomposites polymer/organoclay studies attained very big widespreading [1, 2]. However, the majority of works fulfilled on this theme has mainly an applied character

---

[1] Chernyshevsky st., 173, Nal'chik-360004, Russian Federation.

[2] Kosygin st., 4, Moscow-119334, Russian Federation.

and theoretical aspects of polymers reinforcement by organoclays are studied much less. So, the authors [3] developed for this purpose multiscale micromechanical model, in the basis of which representation about organoclay "effective particle" was assumed. The indicated "effective particle" includes in itself both nanofiller platelets and adjoining to them (or located between them) polymeric matrix layers. Despite the complexity of model [3] it has an essential lacks number. First, the indicated above complexity results to necessity of parameters large number usage, a part of which is difficult enough and sometimes impossible to determine. Secondly, this model is based on micromechanical models application, which in essence exhausted their resources [4]. Thirdly, at the entire of its complexity the model [3] takes not into account such basis for polymer nanocomposites factors as a real level of interfacial adhesion nanofiller-polymeric matrix and polymer chain flexibility for nanocomposite matrix.

The other model [2], used for the same purpose, is based on the following percolation relationship application [5]:

$$\frac{E_n}{E_m} = 1 + 11\varphi_n^{1.7},$$

(1)

where E_n and E_m are elasticity modulus of nanocomposite and matrix polymer, accordingly, φ_n is nanofiller volume contents.

Let us note, that the model [2] is based on the approach, principally different from micromechanical models: it is assumed that polymer composites properties are defined by their matrix structural state only and filler role consists in modification and fixation of matrix polymer structure [6].

Proceeding from the said above, the model [2] correctness checking is the present paper purpose with using of the experimental data for nanocomposites epoxy polymer/Na[+]-montmorillonite (EP/MMT) [7] and also the comparison of models [2] and [3] theoretical estimations for the mentioned nanocomposites.

EXPERIMENTAL

As matrix polymer epoxy polymer on the basis of 3,4-epoxycyclohexylmethyl-3,4-epoxycyclohexane carboxylate (epoxy monomer), cured by hexahydro-4-methylphalic anhydride at molar ratio 0.87-1.0, was used. To this mixture nanofiller Na[+]-montmorillonite under product name Cloisite 30B was added. Beforehand nanofiller was mixed with denatured ethanol for removement of any excess of surfactants from silicate platelets surface. This mixture was then centrifuged at 5000 rpm for 10 min prior to decanting of the ethanol. After Cloisite 30B addition the mixture was stirred in a mixer speedmixer DAC 150 FV at a setting of 2500 rpm for 45 s and then at 3000 rpm during the same time [7].

The nanocomposites EP/MMT samples were prepared in three stages. The samples were first cured isothermally for up to 8 h at temperatures 353, 373, 393 and 413 K, followed by 8 h at 453 K and finally 12 h at 493 K under vacuum. The samples were cooled down at a rate of 1-2 K/min and $20 \times 5 \times 1$ mm^3 pieces for dynamic spectroscopy measurements were machined. Na^+-montmorillonite contents in the studied nanocomposites made up 2, 5, 10 and 15 mass. % [7].

The dynamical mechanical tests were carried out by using Rheometric Scientific ARES rheometer in torsional mode at frequency 1 s^{-1} in nitrogen atmosphere.

The nanocomposites structure studies using wide-angle X-ray diffractometry were carried out on Bruker-AXS General Area Diffraction Detection System using Cu irradiation with wave length $\lambda=1.54$ Å. The distance between Na^+-montmorillonite platelets (d_{001} spacing) was calculated according to these measurements [7].

RESULTS AND DISCUSSION

The model [3] essence consists in the following. Composite continuous simulation found out that properties improvement depends strongly on filler particles individual features: their volume fraction f_g, particle aspect ratio (anisotropy) L/t and ratio of particle and matrix mechanical properties. These important aspects of nanocomposites polymer/organoclay require co-ordinated and precise definition. The authors [3] used multiscale simulation strategy for the calculation of nanocomposite hierarchical morphology at scale of order of thousand microns the large particle aspect ration within the matrix limits represents the structure; at scale of several microns clay particle structure presents itself either fully divided organoclay platelets with thickenss of order of nanometer, or organoclay parallel platelets packing, separated by interlayer galleries with thickness of nanometer level, and polymeric matrix. In this case the quantitative structural parameters, obtained by X-ray diffraction and electron microscopy methods (silicate platelets number N in organoclay bundle, spacing d_{001} between silicate platelets) were used by the authors [3] for definition of clay "particles" geometrical features including L/t and ratio f_p to nanofiller mass content W_n. These geometrical features together with silicate platelet stiffness estimations, obtained from molecular dynamics simulations, give basis for prediction of organoclay particle effective mechanical properties. It is easy to see that it is just such approach that defines the model [3] complexity.

Unlike many mineral fillers, used at plastics production (talcum powder, mica ect.), organoclays, in particular montmorillonite, are capable to stratify and disperse into separate platelets with thickness of about 1 nm [3]. The montmorillonite platelets bundles, in separable after the introduction into polymer, are often called tactoids. The term "intercalation" describes the case, when small polymer amounts penetrate into galleries between silicate platelets, that causes these platelets separation on the value ~ 2-3 nm. Exfoliation or stratification occurs at the distance between platelets (in X-raying this distance is accepted to be called the spacing d_{001}) of order of 8-10 nm. Well stratified and dispersed nanocomposite includes separate organoclay platelets, uniformly distributed in polymeric matrix [1]. The equation (1) usage for the prediction of nanocomposites reinforcement degree E_n/E_m has shown the necessity of its following modification [1, 2]. First of all it has been found out that not only in the full sense nanofiller (organoclay platelets), but also interfacial regions, formed on its platelets surface, with relative fraction φ_{if} are nanocomposites structure reinforcing or strengthening element. Such situation is due to interfacial layers higher stiffness in comparison with bulk polymeric matrix in virtue of strong interactions polymer – organoclay and molecular mobility suppression in the mentioned layers [2]. Thus, Na^+-montmorillonite platelets and formed on their surface interfacial regions also presents themselves organoclay "effective particle" with volume concept ($\varphi_n + \varphi_{if}$), where φ_n is nanofiller volume concept. Then the equation (1) can be rewritten as follows [1]:

$$\frac{E_n}{E_m} = 1 + 11\left(\varphi_n + \varphi_{if}\right)^{1.7}.$$ (2)

Besides, within the frameworks of fractal model of interfacial layers formation in polymer nanocomposites it has been shown that between φ_{if} and φ_n the following relationship exists [8]:

$$\varphi_{if} = 0.955\varphi_n b$$ (3)

for intercalated organoclay and

$$\varphi_{if} = 1.910\varphi_n b$$ (4)

for exfoliated one, where b is parameter, characterizing the nanofiller – polymeric matrix interfacial adhesion level [9]. Let us note that the equations (3) and (4) were obtained by accounting for organoclay platelets strong anisotropy. For example, for particulate (approximately spherical) nanofiller particles the similar equation has the following form [2]:

$$\varphi_{if} = 0.102\varphi_n b.$$ (5)

The equations (2)-(4) combination allows to obtain the final variant of the formulae for nanocomposites polymer/organoclay reinforcement degree determination [2]:

$$\frac{E_n}{E_m} = 1 + 11\left(1.955\varphi_n b\right)^{1.7}$$ (6)

for intercalated organoclay and

$$\frac{E_n}{E_m} = 1 + 11\left(2.910\varphi_n b\right)^{1.7}$$ (7)

for exfoliated one.

Therefore, the equations (2)-(7) comparison demonstrates that the model [2] accounts for nanofiller volume contents, its particles anisotropy degree and interfacial adhesion level, but does not account for nanofiller characteristics (for example, its elasticity modulus), that differs principally this model from the treatment [3].

The nanofiller volume fraction φ_n can be estimated as follows. At it is known [5], the following interconnection exists between nanofiller volume φ_n and mass W_n contents:

$$\varphi_n = \frac{W_n}{\rho_n}, \tag{8}$$

where ρ_n is the nanofiller density, determined according to the equation [5]:

$$\rho_n = \frac{6}{S_u D_p}, \tag{9}$$

where S_u is the filler specific surface, which is equal to $\sim 74\times10^3$ m^2/kg for Na$^+$-montmorillonite [11], D_p is its particles size. Since Na$^+$-montmorillonite particle length is anisotropic and has length is ~ 100 nm, width is ~ 35 nm and thickness is ~ 1 nm [10], then as D_p these sizes arithmetical mean was chosen. Then the value ρ_n=1790 kg/m^3.

In paper [7] it has been found out experimentally that for nanocomposites EP/MMT with W_n=2 and 5 mass. % the interlayer spacing value $d_{001} \geq 10$ nm and for the same nanocomposites with W_n=10 and 15 mass. % d_{001}=6.3-9.8 nm. Therefore, according to the adduced above classification nanocomposites with W_n=2 and 5 mass. % are to be defined as having exfoliated organoclay and with W_n=10 and 15 mass. % – as intercalated organoclay. In Figure 1 the curves $E_n(W_n)$, calculated for the indicated above cases according to the equations (7) and (6), accordingly, at the condition b=1 (perfect adhesion by Kerner) and also the experimental data for nanocomposites EP/MMT (points) are adduced. As one can see, the good correspondence of the data for EP/MMT with W_n=10 and 15 mass. % and calculation according to the equation (6) and with W_n=2 and 5 mass. % – to the equation (7), that is provided for the model [2]. Therefore, the mentioned model reflects well both nanofiller contents and its platelets stratification degree and in addition it is much simpler and physically much clearer than multiscale model [3].

One more aspect of the model [2] concerns of the parameter b determination, which characterizes interfacial adhesion nanofiller – polymeric matrix level. The condition b=1 was accepted for the value E_n estimation (Figure 1), but this parameter can also be estimated more precisely according to the equations (6) and (7), using experimental values E_n, E_m and φ_n independent estimation of interfacial adhesion level for the studied nanocomposites can be obtained with the aid of the parameter A, determined according to the equation [12]:

$$A = \frac{1}{1-\varphi_n}\frac{\operatorname{tg}\delta_n}{\operatorname{tg}\delta_m} - 1, \tag{10}$$

where tg δ_n and tg δ_m are mechanical losses angle tangents for nanocomposite and matrix polymer, accordingly, the values of which are adduced in paper [7].

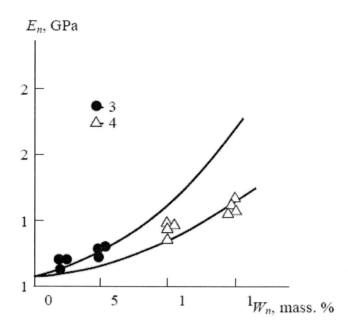

Figure 1. The comparison of calculated according to the equations (6) (1) and (7) (2) and experimental of exfoliated (3) and intercalated (4) organoclay dependences of elasticity modulus En on organoclay mass contents Wn for nanocomposites EP/MMT.

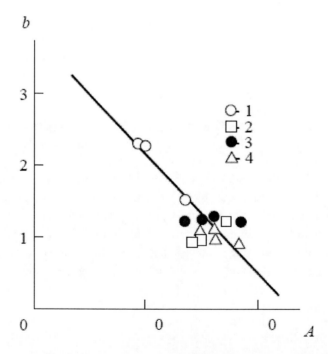

Figure 2. The relation of parameter b and A, characterizing interfacial adhesion level, for nanocomposites EP/MMT with organoclay mass contents W_n=2 (1), 5 (2), 10 (3) and 15 (4) mass. %.

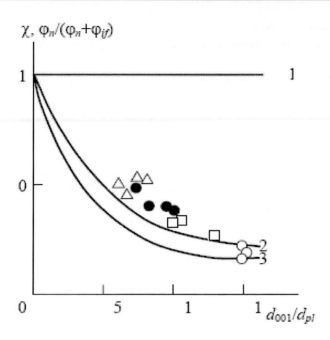

Figure 3. The theoretical dependences of organoclay fraction in "effective particle" χ for silicate platelets number in it N=1 (1), 2 (2) and 4 (3) and parameter $\varphi n/(\varphi n+\varphi if)$ for nanocomposites EP/MMT (points) on the ratio d001/dpl value. Conventional signs are the same that in Figure 2.

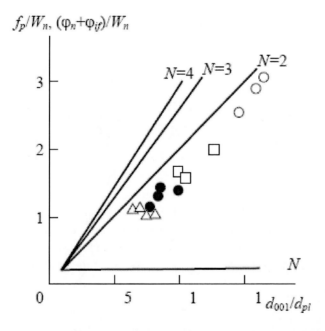

Figure 4. The dependences of organoclay fraction in "effective particles" volume and mass fractions ratio f_p/W_n in the model [3] for silicate platelets number in it N=1 (1), 2 (2), 3 (3) and 4 (4) and $(\varphi_n+\varphi_{if})/W_n$ in the model [2] for nanocomposites EP/MMT (points) on the ratio d_{001}/d_{pl} value. Conventional signs are the same that in Figure 2.

As it is known [12], A decrease means interfacial adhesion level increase. In Figure 2 the comparison of parameters A and b, characterizing interfacial adhesion level is adduced for nanocomposites EP/MMT. As it has been expected, A decrease according to the indicated above reasons corresponds to b growth, i.e., determined according to the equations (6) and (7) parameter b characterizes the real level of interfacial adhesion in nanocomposites polymer/organoclay. As it was assumed above, the greater part of the obtained by the indicated mode values b is close to a unit.

Further let us compare organoclay "effective particle" characteristics in models [2] and [3]. The authors [3] calculated the dependences of silicate relative fraction χ in "effective particle" and plotted its dependences on ratio d_{001}/d_{pl} for different N, where d_{pl} is silicate platelet thickness ($d_{pl}=1$ nm), N is these platelets number in "effective particle". In Figure 3 theoretical calculations according to the model [3] are shown by solid lines for $N=1$, 2 and 4, and experimental dependences of $\varphi_n/(\varphi_n+\varphi_{if})$ in the EP/MMT – by points. As one can see, the similar trends of χ and $\varphi_n/(\varphi_n+\varphi_{if})$ change as a function of d_{001}/d_{pl} are obtained, that indicates the identity of "effective particle" definition in the models [2] and [3]. The comparison of theoretical curves and experimental points in Figure 3 shows that for nanocomposites EP/MMT the value N is somewhat smaller than 2, i.e., organoclay in these nanocomposites is stratified enough.

In Figure 4 theoretical dependences of "effective porticles" volume fraction and nanofiller mass contents ratio f_p/W_n on ratio d_{001}/d_{pl} value, shown by straight lines for $N=1$, 2, 3 and 4, are adduced. In the same Figure the dependences of ratio of sum $(\varphi_n+\varphi_{if})$, which in the model [2] is f_p analogue, and W_n, obtained according to the equations (6) and (7), for nanocomposites EP/MMT (points) are given. The similarity of trends of the indicated ratios as a function of d_{001}/d_{pl} change is observed again, which confirms again "effective particle" definition in the models [2] and [3] identity. And as earlier, for nanocomposites EP/MMT the value N can be estimated as somewhat smaller than $N=2$.

CONCLUSIONS

As the present paper results have shown, the offered by authors [2] the model of organoclay "effective particle" describes nanocomposites polymer/organoclay structure and properties so well, as the multiscale micromechanical model, elaborated in paper [3]. However, unlike latter, the model [2] is much simpler, has clear physical significance, does not require micromechanical models application and, hence, organoclay and interfacial regions characteristics usage, but accounts for interfacial adhesion polymeric matrix – nanofiller real level. The indicated factors make the model [2] more suitable for applied calculations and clearer from the point of view of physical treatment in comparison with the model [3].

REFERENCES

[1]   Malamatov A.Kh., Kozlov G.V., Mikitaev M.A. The Reinforcement Mechanisms of Polymer Nanocomposites. Moscow, Publishers the D.I. Mendeleev RKhTU, 2006, 240 p.

[2]   Mikitaev A.K., Kozlov G.V., Zaikov G.E. Polymer Nanocomposites: Variety of Structural Forms and Applications. New York, Nova Science Publishers, Inc., 2008, 319 p.

[3]   Sheng N., Boyce M.C., Parks D.M., Rutledge G.C., Abes J.I., Cohen R.E. Multiscale micromechanical modeling of polymer/clay nanocomposites and the effective clay particle. // Polymer, 2004, v. 45, № 2, pp. 487-506.

[4]   Ahmed S., Jones F.R. A review of particulate reinforcement theories of polymer composites. // J. Mater. Sci., 1990, v. 25, № 12, pp. 4933-4942.

[5]   Bobryshev A.N., Kozomazov V.N., Babin L.O., Solomatov V.I. Synergetics of Composite Materials. Lipetsk, NPO ORIUS, 1994, 154 p.

[6]   Novikov V.U., Kozlov G.V. A review of particulate reinforcement theories of polymer composites. // Mekhanika Kompozitnykh Materialov, 1999, v. 35, № 3, pp. 269-290.

[7]   Chen J.-S., Poliks M.D., Ober C.K., Zhang Y., Wiesner U., Giannelis E. Study of the interlayer expansion mechanism and thermal-mechanical properties of surface-initiated epoxy nanocomposites. // Polymer, 2002, v. 43, № 14, pp. 4895-4904.

[8]   Kozlov G.V., Malamatov A.Kh., Antipov E.M., Karnet Yu.N., Yanovskii Yu.G. Structure and mechanical properties of polymer nanocomposites within the frameworks of fractal concepts. // Mekhanika Kompozitsionnykh Materialov I Konstruktsii, 2006, v. 12, № 1, pp. 99-140.

[9]   Kozlov G.V., Aphashagova Z.Kh., Burya A.I., Lipatov Yu.S. Nanoadhesion and reinforcement mechanism of particulate-filled polymer nanocomposites. // Inzhenernaya Fizika, 2008, № 1, pp. 47-50.

[10]   Dennis H.R., Hunter D.L., Chang D., Kim S., White J.L., Cho J.W., Paul D.R. Effect of melt processing conditions on extent of exfoliation in organoclay-based nanocomposites. // Polymer, 2001, v. 42, № 24, pp. 9513-9522.

[11]   Pernyeszi T., Dekany I. Surface fractal and structural properties of layered clay minerals monitored by small-angle X-ray scattering and low-temperature nitrogen adsorption experiments. // Colloid Polymer Sci., 2003, v. 281, № 1, pp. 73-78.

[12]   Kubat J., Rigdahl M., Welander M. Characterization of interfacial interactions in high density polyethylene filled with glass spheres using dynamic-mechanical analysis. // J. Appl. Polymer Sci., 1990, v. 39, № 9, pp. 1527-1539.

In: Polymer Yearbook – 2011.
Editors: G. Zaikov, C. Sirghie et al. pp. 147-156

ISBN 978-1-61209-645-2
© 2011 Nova Science Publishers, Inc.

Chapter 14

THE NANOCOMPOSITES POLYETHYLENE/ORGANOCLAY PERMEABILITY TO GAS DESCRIPTION WITHIN THE FRAMEWORKS OF PERCOLATION AND MULTIFRACTAL MODELS

G. V. Kozlov[1], B. Zh. Dzhangurazov.[1],
G. E. Zaikov[2] and A. K. Mikitaev[1]

[1]Kabardino-Balkarian State University,[1]
Nal'chik-360004, Russian Federation
[2]N.M. Emanuel Institute of Biochemical Physics
of Russian Academy of Sciences,[2]
Moscow-119334, Russian Federation

ABSTRACT

It has been shown, that permeability to gas coefficient reduction at layered nanofiller introduction in polyethylene is due to polymer matrix fraction decrease, which is accessible for gas transport processes. Two models (percolation and multifractal ones) are offered for this reduction quantitative description.

Keywords: Nanocomposite, organoclay, permeability to gas, interfacial regions, percolation and multifractal models.

INTRODUCTION

In the last 15 years nanocomposites polymer/organoclay elaboration causes great interest because of their physical and mechanical properties essential improvement in comparison

---

[1] Chernyshevsky st., 173, Nal'chik-360004, Russian Federation.

[2] Kosygin st., 4, Moscow-119334, Russian Federation.

with the initial matrix polymer at small (no more than 10 mass %) nanofiller contents [1, 2]. The essential reduction of permeability to gas coefficient P is one the indicated changes of these nanocomposites properties. So, in a number of papers [3-7] it has been shown, that the introduction in polyethylenes of montmorillonite at volume contents of the last 0.005-0.035 decreases the value P in several times in comparison with the initial matrix polymer. The analysis of this effect, fulfilled by the authors of the indicated papers, assumes, that permeability to gas coefficient reduction is related to essential increase of gas molecules way winding at diffusion through nanocomposite film, containing montmorillonite anisotropic particles, but not to matrix polyethylene structure change at nanofiller (montmorillonite) introduction. As a rule, at this effect description models, were used where aspect ratio of montmorillonite platelet, i.e. its anisotropic level, is the main parameter [4, 8]. However, other models also exist, describing this effect and taking into consideration characteristics both nanocomposite structure and gas-penetrant molecule structure. For example, the multifractal model [9] allows quantitative estimation of gas-penetrant molecules way winding degree through polymer material. Therefore the purpose of the present paper is structural analysis of permeability to gas coefficient reduction effect for nanocomposites low density polyethylene/montmorillonite [7] with percolation and multifractal models using.

EXPERIMENTAL

Low density polyethylene (LDPE) of industrial production mark El-Lene, LD1905F with melt flow index 5 g/10 min and density 0.919 g/cm^3 was used as matrix polymer. Purified Na$^+$-montmorillonite mark Bentolite H (MMT) was used as nanofiller. To prepare organoclay, MMT was mixed with di(hydrogenated tallowalkyl) dimethyl ammonium chloride. Polyethylene grafted maleic anhydride was applied as compatibilizing agent [7].

Nanocomposites LDPE/MMT were prepared by the indicated components mixed in melt on twin-screw extruder Thermo Haake Rheomex RTW at temperature 433 K and screw speed of 150 rpm [7].

Tensile tests were performed by using a mechanical testing machine Instron, model 5567, at temperature 293 K and strain rate ~ 2×10^{-2} s^{-1}. For each test, 10 film specimens (50-60 µm thickness) were used [7].

The permeability to gas coefficient measurements for initial LDPE and nanocomposites LDPE/MMT were performed on apparatus Oxygen Permeation Analyzer, model 8501 with automatical recording of the indicated parameter [7].

RESULTS AND DISCUSSION

The authors [10] proposed the percolation relationship for polymer composites reinforcement degree E_c/E_m description:

$$\frac{E_c}{E_m} = 1 + 11\varphi_n^{1.7},$$

(1)

where E_c and E_m are elasticity moduli of composite and matrix polymer, accordingly, φ_n is filler volume contents.

For polymer nanocomposites the equation (1) the following modification was proposed [11]:

$$\frac{E_n}{E_m} = 1 + 11\left(\varphi_n + \varphi_{if}\right)^{1.7},$$
(2)

where E_n is nanocomposite elasticity modulus, φ_{if} is interfacial regions relative fraction.

The nanofiller volume fraction φ_n was determined according to the known equation [11]:

$$\varphi_n = \frac{W_n}{\rho_n},$$
(3)

where W_n is nanofiller mass contents, ρ_n is montmorillonite density, approximately equal to 2000 kg/m^3 [4].

As a rule, nanofiller contents φ_n (or W_n) increase is accompanied by E_n enhancement and permeability to gas coefficient P_n reduction of nanocomposites in comparison with similar parameter for matrix polymer P_m. Since at the deduction of the equation (1) derivation regulations were applied, which are true in gas transport processes case as well (nanofiller critical concentration absence, much lower nanofiller permeability to gas coefficient in comparison with P_m), then the equation (2) can be used for permeability to gas coefficient relative change description in the following form:

$$\frac{P_m}{P_n} = 1 + 11\left(\varphi_n + \varphi_{if}\right)^{1.7},$$
(4)

where the total value $(\varphi_n+\varphi_{if})$ can be obtained by mechanical testing results according to the equation (2)

The equations (2) and (4) comparison shows, that E_n/E_m increase should be accompanied by P_m/P_n enhancement or P_n reduction. In Figure 1 the relation between reinforcement degree E_n/E_m and relative permeability to gas coefficient P_m/P_n is adduced, which turns out to be linear with a slope close to one. However, this correlation does not pass through coordinates origin and at $E_n/E_m \approx 1.22$ the value $P_m/P_n=1$, i.e. at MMT small contents the value P_n is close to permeability to gas coefficient for matrix polymer. The indicated circumstance makes it necessary the equation (4) modification as follows:

$$\frac{P_m}{P_n} = 0.78 + 11\left(\varphi_n + \varphi_{if}\right)^{1.7}.$$
(5)

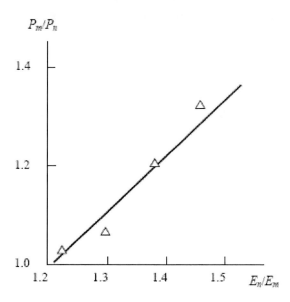

Figure 1.The comparison of relative permeability to gas coefficient P_m/P_n and reinforcement degree E_n/E_m for nanocomposites LDPE/MMT.

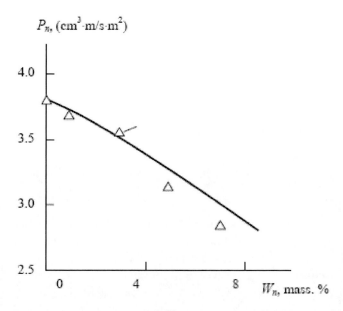

Figure 2. The comparison of the experimental (1) and calculated according to the equation (5) (2) dependences of permeability to gas coefficient P_n on organoclay mass contents W_n for nanocomposites LDPE/MMT.

The comparison of the experimental and calculated according to the equation (5) (percolation model) dependences of permeability to gas coefficient P_n as a function of nanofiller mass contents W_n for nanocomposites LDPE/MMT is shown in Figure 2. As it follows from this comparison, the equation (5) gives enough precise description of permeability to gas coefficient for the studied nanocomposites at nanofiller contents growth (the theory and experiment average discrepancy makes up 3 %).

Another variant of nanocomposites permeability to gas coefficient P_n calculation assumes gas transport multifractal model [9] using. The indicated model allows to determine the value P_n as follows:

$$P_n = P_m \alpha_{ac} D_{ch},$$ (6)

where α_{ac} is polymer nanocomposite relative fraction, accessible for gas transport processes, D_{ch} is fractal dimension of polymer chain section between its fixation points (chemical cross-linking nodes, physical entanglements, clusters, etc.), which characterizes molecular mobility level in polymer [9].

The authors [12] proposed the following equation for semicrystalline polymers permeability to gas coefficient estimation:

$$P = \frac{P_{am}}{\tau \beta},$$ (7)

where P_{am} is permeability to gas coefficient of completely amorphous polymer, τ is nonlinearity (winding) coefficient, which is due to difficulty of gas-penetrant molecules transport ways between crystallites, β is the so-called polymer chains immobility coefficient.

The direct similarity between nanocomposites and semicrystalline polymers consists in the fact that both indicated polymer materials classes have regions, impenetrable for transport processes (nanofiller and cristallites, accordingly). Therefore the indicated processes are realized through polymer matrix and amorphous phase, respectively. Hence, from the relationships (6) and (7) can be written:

$$P_m \sim P_{am},$$ (8)

$$\tau = \frac{1}{\alpha_{ac}},$$ (9)

$$\beta = \frac{1}{D_{ch}}.$$ (10)

Thus, the offered treatment assumes in the general case not only nonlinearity coefficient τ change because of nanofiller introduction, but polymer matrix structure change in virtue of coefficient β variation, depending on its molecular mobility level [9].

Let us consider estimation methods of parameters, included in the equation (6). The value P_m is accepted equal to value P for LDPE. As it is known [11], in the polymer nanocomposites case actually nanofiller and surrounded its particles interfacial regions with relative fractions φ_n and φ_{if}, accordingly (see also the equation (5)) will be impenetrable for gas transport processes. Then the polymer matrix relative fraction α can be estimated as follows:

$$\alpha = 1 - \varphi_n - \varphi_{if}. \tag{11}$$

In its turn, the following relation between values φ_n and φ_{if} exists for the layered silicates [11]:

$$\varphi_{if} = 1.910\varphi_n \tag{12}$$

for exfoliated organoclay and

$$\varphi_{if} = 0.955\varphi_n \tag{13}$$

for an intercalated one.

The nanocomposites polymer matrix fraction, accessible for oxygen permeability, α_{ac} is determined according to the equation [9]:

$$\alpha_{ac} = \alpha^{d_{O_2}}, \tag{14}$$

where d_{O_2} is oxygen molecule diameter, which is equal to 2.9 Å [13].

Dimension D_{ch} can be determined with the aid of the following equation [14]:

$$\frac{2}{\varphi_{cl}} = C_\infty^{D_{ch}}, \tag{15}$$

where φ_{cl} is polymer matrix local order domains (clusters) relative fraction, C_∞ is characteristic ratio, which is polymer chain statistical flexibility indicator [15].

For parameters φ_{cl} and C_∞ estimation it is necessary to determine the fractal (Hausdorff) dimension of nanocomposites structure d_f, that can be made according to the equation [16]:

$$d_f = (d-1)(1+\nu), \tag{16}$$

where d is dimension of Euclidean space, in which fractal is considered (it is obvious, that in our case $d=3$), ν is Poisson's ration, estimated according to mechanical tests results with the aid the relationship [17]:

$$\frac{\sigma_Y}{E_n} = \frac{1-2\nu}{6(1+\nu)}, \tag{17}$$

where σ_Y is nanocomposites yield stress.

Further the value C_∞ can be calculated according to the equation [14]:

$$C_\infty = \frac{2d_f}{d(d-1)(d-d_f)} + \frac{4}{3}.$$

(18)

Then value φ_{cl} was determined, by using the following formula [14]:

$$d_f = 3 - 6\left(\frac{\varphi_{cl}}{C_\infty S}\right)^{1/2},$$

(19)

where S is macromolecule cross-sectional area, which is equal to 14.4 Å^2 for polyethylenes [18].

The crystallinity degree K of the studied nanocomposites can be estimated according to the equation [19]:

$$K = 0.32 C_\infty^{1/3}.$$

(20)

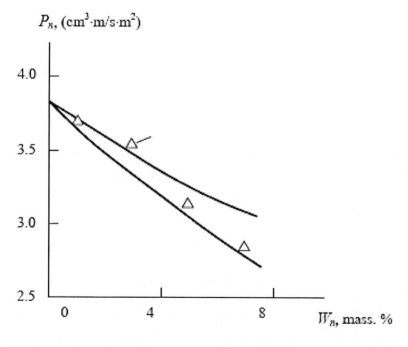

Figure 3.The comparison of the experimental (1) and calculated according to the equation (21) (2, 3) in supposition of intercalated (2) and exfoliated (3) nanofiller dependences of permeability to gas coefficient P_n on organoclay mass contents W_n for nanocomposites LDPE/MMT.

As estimations according to the equations (15) and (20) have shown, the values D_{ch} and K change insignificantly (D_{ch}=1.73-1.75, K=0.49-0.51), therefore for the studied nano-composites the equation (6) is simplified up to:

$$P_n = P_m \left(1 - \varphi_n - \varphi_{if}\right)^{d_{O_2}} .$$

(21)

In Figure 3 the comparison of the experimental and calculated according to the equation (21) (multifractal model) in supposition of both exfoliated and intercalated layered silicate dependences $P_n(W_n)$ for nanocomposites LDPE/MMT is adduced. As one can see, the miltifractal model of gas transport processes also gives a good correspondence with experiment (the theory and experiment average discrepancy makes up 3 %).

CONCLUSIONS

The layered nanofiller introduction in low density polyethylene results to the received nanocomposites permeability to gas coefficient reduction. The fulfilled structural analysis of this effect demonstrated, that the indicated reduction was due to decrease of polymer matrix relative fraction, accessible for gas transport process, in virtue of both nanofiller introduction and interfacial regions formation. Both used theoretical models (percolation and multifractal ones) give correct quantitative description of permeability to gas coefficient reduction at nanofiller contents increasing.

REFERENCES

[1]   Giannelis E.P., Krishnamoorti R., Manias E. *Adv. Polymer Sci.,* 1999, v. 138, № 1, p. 107-147.
[2]   LeBaron P.C., Wang Z., Pinnivaia T.J. *Appl. Clay Sci.,* 1999, v. 15, № 1, p. 11-29.
[3]   Kovaleva N.Yu., Brevnov P.N., Grinev V.G., Kuznetsov S.P., Pozdnyakova I.V., Chvalun S.N., Sinevich E.A., Novokshonova L.A. *Vysokomolek. Soed.* A, 2004, v. 46, № 6, p. 1045-1051.
[4]   Hotta S., Paul D.R. Polymer, 2004, v. 45, № 21, p. 7639-7654.
[5]   Durmus A., Woo M., Kasgöz A., Macosko C.W., Tsapatsis M. *Eur. Polymer J.,* 2007, v. 43, № 14, p. 3737-3749.
[6]   Reddy M.M., Gupta R.K., Bhattacharaya S.N., Parthasarathy R. Korea-Australia Rheology J., 2007, v. 19, № 3, p. 133-139.
[7]   Arunvisut S., Phummanee S., Somwangthanaroj A. *J. Appl. Polymer Sci.,* 2007, v. 106, № 6, p. 2210-2217.
[8]   Utracki L.A., Lyngaae-Jorgensen *J. Rheol. Acta,* 2002, v. 41, № 3, p. 394-407.
[9]   Kozlov G.V., Zaikov G.E., Mikitaev A.K. The Fractal Analysis of Gas Transport in Polymers: the Theory and Practical Applications. New York, Nova Science Publishers, Inc., 2009, 238 p.

[10] Bobryshev A.N., Kozomazov V.N., Babin L.O., Solomatov V.I. Synergetics of Composite Materials. Lipetsk, NPO ORIUS, 1994, 154 p.

[11] Mikitaev A.K., Kozlov G.V., Zaikov G.E. Polymer Nanocomposites: the Variety of Structural Forms and Applications. New York, Nova Science Publishers, Inc., 2008, 319 p.

[12] Ash R., Barrer R.M., Palmer D.G. Polymer, 1970, v. 11, № 8, p. 421-430.

[13] Teplyakov V.V., Durgar'yan S.G. *Vysokomolek. Soed. A*, 1984, v. 24, № 7, p. 1498-1505.

[14] Kozlov G.V., Zaikov G.E. Structure of the Polymer Amorphous State. Utrecht, Boston, Brill Academic Publishers, 2004, 465 p.

[15] Budtov V.P. Physical Chemistry of Polymer Solutions. Sankt-Peterburg, Khimiya, 1992, 384 p.

[16] Balankin A.S. Synergetics of Deformable Body. Moscow, Publishers of Ministry of Defence SSSR, 1991, 404 p.

[17] Kozlov G.V., Sanditov D.S. Anharmonic Effects and Physical-Mechanical Properties of Polymers. Novosibirsk, Nauka, 1994, 261 p.

[18] Aharoni S.M. Macromolecules, 1985, v. 18, № 12, p. 2624-2630.

[19] Aloev V.Z., Kozlov G.V. The Physics of Orientational Phenomena in Polymer Materials. Nal'chik, Polygraphservice and T, 2002, 288 p.

In: Polymer Yearbook – 2011.
Editors: G. Zaikov, C. Sirghie et al. pp. 157-165

ISBN 978-1-61209-645-2
© 2011 Nova Science Publishers, Inc.

Chapter 15

THE RHEOLOGY OF PARTICULATE-FILLED POLYMER NANOCOMPOSITES

G. V. Kozlov[1], M. A. Tlenkopachev [2] and G. E. Zaikov [3]

[1])Kabardino-Balkarian State University,[1]
Nal'chik-360004, Russian Federation
[2])Material Research Institute of National Autonomous University of Mexico
70-360, CU, Coyoacan, Mexico DF, 04510, Mexico
[3])N.M. Emanuel Institute of Biochemical Physics
of Russian Academy of Sciences, [2]
Moscow-119334, Russian Federation

ABSTRACT

It has been found out that continuous models do not give adequate description of melt viscosity for particulate-filled polymer nanocomposites. The correct treatment of the mentioned viscosity can be obtained within the frameworks of viscous liquid flow fractal model. It has been shown that such approach differs principally from the used ones at microcomposites viscosity description.

Keywords: Nanocomposite, disperse particles, melt, viscosity, aggregation, fractal analysis.

INTRODUCTION

An inorganic nanofiller of various types usage for polymer nanocomposites production has been widely spread [1]. However, the mentioned nanomaterials melt properties are not studied completely enough. As a rule, when nanofillers application is considered, then

[1] Chernyshevsky st., 173, Nal'chik-360004, Russian Federation.

[2] Kosygin st., 4, Moscow-119334, Russian Federation.

compromise between mechanical properties in solid state improvement, melt viscosity at processing enhancement, nanofillers dispersion problem and process economic characteristics is achieved. Proceeding from this, the relation between nanofiller concentration and geometry and nanocomposites melt properties is an important aspect of polymer nanocomposites study. Therefore the purpose of the present paper is an investigation and theoretical description of the dependence of nanocomposite high density polyethylene/calcium carbonate [2] melt viscosity on nanofiller concentration.

EXPERIMENTAL

High density polyethylene (HDPE) of industrial production mark 276-73 was used as matrix polymer and nanodimensional calcium carbonate ($CaCO_3$) in the form of compound mark Nano-Cal NC-K0117 (China) with particles size of 80 nm and mass contents 1-10 mass % was used as nanofiller.

Nanocomposites HDPE/ $CaCO_3$ were prepared by components mixing in melt on twin-screw extruder Thermo Haake, model Reomex RTW 25/42, production of German Federal Republic. Mixing was performed at temperature 483-493 K and screw speed of 15-25 rpm during 5 min. Testing samples were obtained by casting under pressure method on casting machine Test Samples Molding Apparate RR/TS MP of firm Ray-Ran (Taiwan) at temperature 473 K and pressure 8 MPa.

Nanocomposites viscosity was characterized by melt flow index (MFI). The measurements of MFI were performed on extrusion-type plastometer with capillary diameter of 2.095±0.005, model IIRT-5, at temperature 513 K and load of 2.16 kg. This sample was maintained at the indicated temperature during 4.5±0.5 min.

Uniaxial tension mechanical tests were performed on samples in the form of two-sided spade with sizes according to GOST 11265-80. Tests were conducted on universal testing apparatus Gotech Testing Machine CT-TCS 2000, production of German Federal Republic, at temperature 293 K and strain rate ~ 2×10^{-3} s^{-1} [2].

RESULTS AND DISCUSSION

For polymer microcomposites, i.e. composites with filler of micron sizes, two simple relations between melt viscosity η, shear modulus G in solid-phase state and filling volume degree φ_n were obtained [3]. The relationship between η and G has the following form:

$$\frac{\eta}{\eta_0} = \frac{G}{G_0},$$

(1)

where η_0 and G_0 are melt viscosity and shear modulus of matrix polymer, accordingly.

Besides, microcomposite melt viscosity increase can be estimated as follows (for $\varphi_n < 0.40$) [3]:

$$\frac{\eta}{\eta_0} = 1 + \varphi_n. \tag{2}$$

In Figure 1 the dependences of ratios G_n/G_m and η_n/η_m, where G_n and η_n are shear modulus and melt viscosity of nanocomposite, G_m and η_m are the same characteristics for the initial matrix polymer, on $CaCO_3$ mass contents W_n for nanocomposites HDPE/$CaCO_3$. Shear modulus G was calculated according to the following general relationship [4]:

$$G = \frac{E}{d_f}, \tag{3}$$

where E is Young's modulus, d_f is nanocomposite structure fractal dimension, determined according to the equation [4]:

$$d_f = (d-1)(1+\nu), \tag{4}$$

where d is dimension of Euclidean space, in which a fractal was considered (it is obvious, that in our case $d=3$), ν is Poisson's ratio, estimated by mechanical tests results with the aid of the relationship [5]:

$$\frac{\sigma_Y}{E} = \frac{1-2\nu}{6(1+\nu)}. \tag{5}$$

MFI reciprocal value was accepted as melt viscosity η measure. The data of Figure 1 clearly demonstrate, that in the case of the studied nanocomposites the relationship (1) is not fulfilled either qualitatively or quantitatively: the ratio η_n/η_m decay at W_n growth corresponds to G_n/G_m enhancement and η_n/η_m absolute values are much larger than the corresponding G_n/G_m magnitudes.

In Figure 2 the comparison of parameters η_n/η_m and $(1+\varphi_n)$ for nanocomposites HDPE/$CaCO_3$ is adduced. The discrepancy between the experimental data and the relationship (2) is obtained again: the absolute values η_n/η_m and $(1+\varphi_n)$ discrepancy is observed and $(1+\varphi_n)$ enhancement corresponds to melt relative viscosity reduction. At the plot of Figure 2 graphing the nominal φ_n value was used, which does not take into consideration nanofiller particles aggregation and estimated according to the equation [6]:

$$\varphi_n = \frac{W_n}{\rho_n}, \tag{6}$$

where ρ_n is nanofiller particles density, which was determined according to the formula [1]:

$$\rho_n = 0.188\left(D_p\right)^{1/3},$$ (7)

where D_p is $CaCO_3$ initial particles diameter.

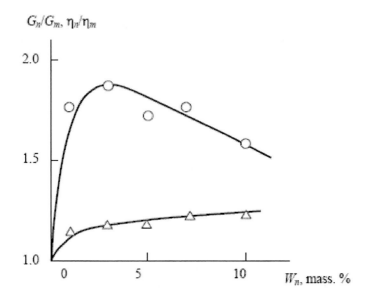

Figure 1.The dependences of shear moduli Gn/Gm (1) and melt viscosities ηn/ηm (2) ratios of nanocomposite Gn, ηn and matrix polymer Gm, ηm on nanofiller mass contents Wn for nanocomposites HDPE/CaCO3.

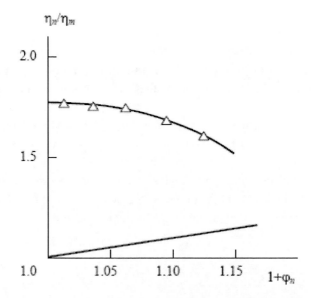

Figure 2.The dependence of nanocomposite and matrix polymer melt viscosities ratio ηn/ηm on nanofiller volume contents (1+φn) (1) for nanocomposites HDPE/CaCO3. The straight line 2 shows relation 1:1.

Hence, the data of Figs. 1 and 2 showed, that the relationships (1) and (2) fulfilled in case of polymer microcomposites are incorrect for nanocomposites. Let us consider this question in more detail. In case of the relationship (1) correctness and Kerner's equation application for G calculation viscosity η_n lower boundary can be obtained according to the equation [3]:

$$\frac{\eta_n}{\eta_m} = 1 + \frac{2.5\varphi_n}{1-\varphi_n}. \tag{8}$$

Since η value is inversely proportional to MFI, then in such treatment the equation (8) can be rewritten as follows:

$$\frac{MFI_m}{MFI_n} = 1 + \frac{2.5\varphi_n}{1-\varphi_n}, \tag{9}$$

where MFI_m and MFI_n are MFI values for matrix polymer and nanocomposite, accordingly.

Three methods can be used for the value φ_n estimation in the equations (8) and (9). The first of them was described above, which gives nominal value φ_n. The second method is usually applied for microcomposites, when massive filler density is used as ρ_n, i.e. ρ_n=const$\approx$2000 kg/m^3 in case of CaCO$_3$. And at last, the third method also uses the equations (6) and (7), but it takes into consideration nanofiller particles aggregation and in this case in the equation (7) initial nanofiller particles diameter D_p is replaced to such particles aggregate diameter D_{ag}. To estimate CaCO$_3$ nanoparticles aggregation degree and, hence, D_{ag} value can be estimated within the frameworks of the strength dispersive theory [7], where yield stress at shear τ_n of nanocomposite is determined as follows:

$$\tau_n = \tau_m + \frac{G_n b_B}{\lambda}, \tag{10}$$

where τ_m is yield stress at shear of polymer matrix, b_B is Burgers vector, λ is distance between nanofiller particles.

In case of nanofiller particles aggregation the equation (10) assumes the look [7]:

$$\tau_n = \tau_m + \frac{G_n b_B}{k(\rho)\lambda}, \tag{11}$$

where $k(\rho)$ is an aggregation parameter.

The included in the equations (10) and (11) parameters are determined as follows. The general relation between normal stress σ and shear stress τ assumes the look [8]:

$$\tau = \frac{\sigma}{\sqrt{3}}.$$
(12)

Burgers vector value b_B for polymer materials is determined from the relationship [9]:

$$b_B = \left(\frac{60.5}{C_\infty}\right)^{1/2}, \text{ Å,}$$
(13)

where C_∞ is characteristic ration, connected with d_f by the equation [9]:

$$C_\infty = \frac{2d_f}{d(d-1)(d-d_f)} + \frac{4}{3}.$$
(14)

And at last, the distance λ between nonaggregated nanofiller particles is determined according to the equation [7]:

$$\lambda = \left[\left(\frac{4\pi}{3\varphi_n}\right)^{1/3} - 2\right]\frac{D_p}{2}.$$
(15)

From the equations (11) and (15) $k(\rho)$ growth follows from 5.5 up to 11.8 in the range of W_n=1-10 mass % for the studied nanocomposites. Let us consider, now such $k(\rho)$ growth is reflected on nanofiller particles aggregates diameter D_{ag}. The equations (6), (7) and (15) combination gives the following expression:

$$k(\rho)\lambda = \left[\left(\frac{0.251\pi D_{ag}^{1/3}}{W_n}\right)^{1/3} - 2\right]\frac{D_{ag}}{2},$$
(16)

allowing at replacement of D_p on D_{ag} to determine real, i.e. with accounting of nanofiller particles aggregation, nanoparticles $CaCO_3$ aggregates diameter. Calculation according to the equation (16) shows D_{ag} increase (corresponding to $k(\rho)$ growth) from 320 up to 580 nm in the indicated W_n range. Further the real value ρ_n for aggregated nanofiller can be calculated according to the equation (7) and real filling degree φ_n – according to the equation (6). In Figure 3 the comparison of the dependences $MFI_n(W_n)$, obtained experimentally and calculated according to the equation (9) with the usage of the values φ_n, estimated by three indicated above methods. As one can see, the obtained according to the equation (9) theoretical results correspond to the experimental data neither qualitatively nor quantitatively.

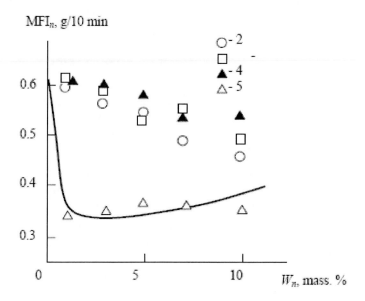

Figure 3. The dependences of melt flow index MFIn on nanofiller mass contents Wn for nanocomposites HDPE/CaCO3. 1 – experimental data, 2-4 – calculation according to the equation (9) without appreciation (2), with appreciation (3) of nanofiller particles aggregation and at the condition ρn=const (4), 5 – calculation according to the relationship (17).

The indicated discrepancy requires the application of principally differing approach at polymer nanocomposites melt viscosity description. Such approach can be fractal analysis, within the frameworks of which the authors [10] were offered the following relationship for fractal liquid viscosity η estimation:

$$\eta(l) \sim \eta_0 l^{2-d_f}, \qquad (17)$$

where l is characteristic linear scale of flow, η_0 is constant, d_f is fractal dimension.

In the considered case the nanoparticles $CaCO_3$ aggregate radius $D_{ag}/2$ follows to the accepted as l. Since the indicated aggregate surface comes into contact with polymer, then its fractal dimension d_{surf} was chosen as d_f. The indicated dimension can be calculated as follows [1]. The value of nanofiller particles aggregate specific surface S_u was estimated according to the equation [11]:

$$S_u = \frac{6}{\rho_n D_{ag}}, \qquad (18)$$

and then the dimension d_{surf} was calculated with the aid of the equation [1]:

$$S_u = 410 \left(\frac{D_{ag}}{2} \right)^{d_{surf} - d} .$$
(19)

As earlier, the value η was considered as reciprocal value of MFI_n and constant η_0 was accepted equal to $(MFI_m)^{-1}$. At these conditions and replacement of proportionality sign in the relationship (17) on equality sign the theoretical values MFI_n can be calculated, if D_{ag} magnitude is expessed in microns. In Figure 3 the comparison of the received by the indicated mode MFI_n values with the experimental dependence $MFI_n(W_n)$, from which theory and experiment good correspondence follows.

The relationship (17) allows to make a number of conclusions. So, at the mentioned above conditions conservation D_{ag} increase, i.e. initial nanoparticles aggregation intensification, results to nanocomposite melt viscosity reduction, whereas d_{surf} enhancement, i.e. nanoparticles surface roughness degree increasing, raises melt viscosity. At d_{surf}=2.0, i.e. nanofiller particles smooth surface, melts viscosity for matrix polymer and nanocomposite will be equal. It is interesting that extrapolation of the MFI_n dependence, obtained experimentally, on the calculated according to the equation (19) d_{surf} values gives the value MFI_n=0.602 g/10 min at d_{surf}=2.0, that is practically equal to the experimental magnitude MFI_m=0.622 g/10 min. The indicated factors, critical ones for nanocomposites, are not taken into consideration in continuous treatment of melt viscosity for polymer composites (the equation (8)).

CONCLUSIONS

The obtained in the present paper results have shown, that the elaborated for microcomposites rheology description models do not give adequate treatment of melt viscosity for particulate-filled nanocomposites. The indicated nanocomposites rheological properties correct description can be obtained within the frameworks of viscous liquid flow fractal model. It is significant, that such approach differs principally from the used ones at microcomposites description. So, nanofiller particles aggregation reduces both melt viscosity and elasticity modulus of nanocomposites in solid-phase state. For microcomposites melt viscosity enhancement is accompanied by elasticity modulus increase.

REFERENCES

[1]   Mikitaev A.K., Kozlov G.V., Zaikov G.E. Polymer Nanocomposites: the Variety of Structural Forms and Applications. New York, Nova Science Publishers, Inc., 2008, 319 p.

[2]   Sultonov N.Zh., Dzhangurazov B.Zh., Mikitaev A.K. Proceedings of IV Internat. Sci.-Pract. Conf. "New Polymer Composite Materials". Nal'chik, KBSU, 2008, p. 285-288.

[3]   Mills N.J. *J. Appl. Polymer Sci.,* 1971, v. 15, № 11, p. 2791-2805.

[4]   Balankin A.S. Synergetics of Deformable Body. Moscow, Publishers Ministry of Defence SSSR, 1991, 404 p.

[5]   Kozlov G.V., Sanditov D.S. Anharmonic Effects and Physical-Mechanical Properties of Polymers. Novosibirsk, Nauka, 1994, 261 p.

[6]   Sheng N., Boyce M.C., Parks D.M., Rutledge G.C., Abes J.I., Cohen R.E. *Polymer,* 2004, v. 45, № 2, p. 487-506.

[7]   Sumita M., Tsukumo Y., Miyasaka K., Ishikawa K. *J. Mater. Sci.,* 1983, v. 18, № 5, p. 1758-1764.

[8]   Honeycombe R.W.K. The Plastic Deformation of Metals. London, Edward Arnold Publishers, LTD, 1968, 402 p.

[9]   Kozlov G.V., Zaikov G.E. Structure of the Polymer Amorphous State. Utrecht, Boston, Brill Academic Publishers, 2004, 465 p.

[10]   Gol'dstein R.V., Mosolov A.B. Doklady RAN, 1992, v. 324, № 3, p. 576-581.

[11]   Bobryshev A.N., Kozomazov V.N., Babin L.O., Solomatov V.I. Synergetics of Composite Materials. Lipetsk, NPO ORIUS, 1994, 154 p.

In: Polymer Yearbook – 2011.   ISBN 978-1-61209-645-2
Editors: G. Zaikov, C. Sirghie et al. pp. 167-174   © 2011 Nova Science Publishers, Inc.

Chapter 16

THE EXPERIMENTAL AND THEORETICAL ESTIMATION OF INTERFACIAL LAYER THICKNESS IN ELASTOMERIC NANOCOMPOSITES

G. V. Kozlov[1), Yu. G. Yanovskii [2) and G. E. Zaikov [3)

[1)Kabardino-Balkarian State University,[1]
Nal'chik-360004, Russian Federation
[2)Institute of Applied Mechanics of Russian Academy of Sciences[2],
Moscow-119991, Russian Federation
[3)N.M. Emanuel Institute of Biochemical Physics
of Russian Academy of Sciences, [3]
Moscow-119334, Russian Federation

ABSTRACT

The interfacial layer thickness and its elasticity modulus were determined experimentally for particulate-filled polymer nanocomposite. It has been found out, that elasticity modulus of interfacial layer exceeds in 5 times corresponding characteristic for bulk polymer matrix. It has been shown, that the theoretical calculation of interfacial layer thickness within the frameworks of fractal model corresponds well to experimental data.

Keywords: Nanocomposite, nanoparticle, interfacial layer, elasticity modulus, nanoscopic method.

---

[1] Chernyshevsky st., 173, Nal'chik-360004, Russian Federation.

[2] Leninskii pr., 32 A, Moscow-119991, Russian Federation.

[3] Kosygin st., 4, Moscow-119334, Russian Federation.

INTRODUCTION

The determination of structural components quantitative characteristics represents an important task, without solution of which quantitative description and prediction of polymer nanocomposites properties cannot be fulfilled. In this aspect interfacial regions play a particular role, since it has been shown earlier, that they are the same reinforcing element in elastomeric nanocomposites as actually nanofiller [1]. Therefore the knowledge of interfacial layer dimensional characteristics is necessary for quantitative determination of one of the most important parameters of polymer composites in general – their reinforcement degree. Proceeding from the said above, the purpose of the present paper is interfacial layer thickness experimental determination with the aid of modern nanoscopic methods [2] and its theoretical calculation within the frameworks of fractal analysis [3].

EXPERIMENTAL

The elastomeric particulate-filled nanocomposite on the basis of butadiene-styrene rubber was an object of the study. Mineral shungite nanodimensional particles (particles average size makes up 40 nm) were used as a nanofiller. The nanoshungite content makes up 37 mass %. Nanodimensional disperse shungite particles were prepared from industrially outputted material by original technology processing. The analysis of the received in milling process shundite particles was monitored with the aid of analytical disk centrifuge (CPS Instruments, Inc., USA), allowing to determine with high precision size and distribution by sizes within the range from 2 nm up to 50 mcm.

Nanostructure was studied on atomic-power microscopes Nano-DST (Pacific Nano-technology, USA) and Easy Scan DFM (Nanosurf, Switzerland) by semi-contact method in the force modulation regime. Atomic-power microscopy results were processed with the aid of specialized software package SPIP (Scanning Probe Image Processor, Denmark). SPIP is powerful programmes package for processing of images, obtained on SPM, AFM, STM, scanning electron microscopes, transmission electron microscopes, interferometers, confocal microscopes, profilometers, optical microscopes and so on. The given package possesses the whole functions number, which are necessary at images precise analysis, in number of which the following are included:

1) the possibility of three-dimensional reflecting objects obtaining, distortions automatized leveling, including Z-error mistakes removal for examination separate elements and so on;
2) quantitative analysis of particles or grains, more than 40 parameters can be calculated for each found particle or pore: area, perimeter, average diameter, the ratio of linear sizes of grain width to its height distance between grains, coordinates of grain center of mass a.a. can be presented in a diagram form or in a histogram form.

The tests by elastomeric nanocomposites nanomechanical properties were carried out by nanoindentation method [2] on apparatus NanoTest 600 (Micro Materials, Great Britain) in loades wide range from 0.01 mN up to 2.0 mN. Sample indentation was conducted in 10 points with interval of 30 mcm. The load was increased with constant rate up to the greatest given load reaching (for the rate 0.05 mN/s-1 mN). The indentation rate was changed in conformity with the greatest load value counting, that loading cycle should take 20 s. The unloading was conducted with the same rate as loading. In the given experiment the "Berkovich's indentor" was used with angle at tip of 65.3° and rounding radius of 200 nm. Indentations were carried out in the checked load regime with preload of 0.001 mN.

For elasticity modulus calculation the obtained in the experiment by nanoindentation course dependences of load on indentation depth (strain) in ten points for each sample at loads of 0.01, 0.02, 0.03, 0.05, 0.10, 0.50, 1.0 and 2.0 mN were processed according to Oliver-Pharr method [4].

RESULTS AND DISCUSSION

In Figure 1 the obtained according to the original methodics results of elasticity moduli calculation for nanocomposite butadiene-styrene rubber/nanoshungite components (matrix, nanofiller particle and interfacial layers), received in interpolation process of nanoindentation data, are presented. The processed in SPIP polymer nanocomposite image with shungite nanoparticles allows experimental determination of interfacial layer thickness l_{if}, which is presented in Figure 1 as steps on elastomeric matrix-nanofiller boundary. The measurements of 34 such steps (interfacial layers) width on processed in SPIP images of interfacial layer various section gave the average experimental value l_{if}=8.7 nm. Besides, nanoindentation results (Figure 1, figure on the right) showed, that interfacial layers elasticity modulus was only by 23-45 % lower than nanofiller elasticity modulus, but it is higher than the corresponding parameter of polymer matrix in 6.0-8.5 times. These experimental data confirm, that for the studied nanocomposite interfacial layer is reinforcing element to the same extent, as actually nanofiller [1, 3, 5].

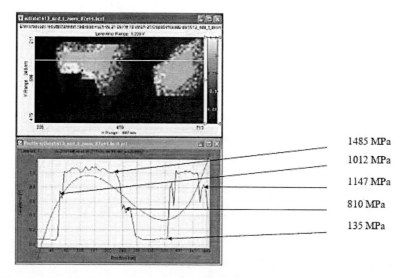

Figure 1. The processed in SPIP image of nanocomposite butadiene-styrene rubber/nanoshungite, obtained by force modulation method, and mechanical characteristics of structural components by the data of nanoindentation (strain 150 nm).

Further let us fulfil the value l_{if} theoretical estimation according to the two methods and compare these results with the ones obtained experimentally. The first method simulates interfacial layer in polymer composites as a result of interaction of two fractals – polymer matrix and nanofiller surface [6, 7]. In this case there is a sole linear scale l, which defines these fractals interpenetration distance [8]. Since nanofiller elasticity modulus is essentially higher, than the corresponding parameter for rubber (in the considered case – in 11 times, see Figure 1), then the indicated interaction reduces to nanofiller indentation in polymer matrix and then $l=l_{if}$. In this case it can be written [8]:

$$l_{if} \approx a\left(\frac{R_p}{a}\right)^{2(d-d_{surf})/d}, \qquad (1)$$

where a is lower linear scale of fractal behaviour, which is accepted equal to statistical segment length l_{st} [9], R_p is nanofiller particle (more precisely, particles aggregate) radius, which for nanoshungite is equal to 167.5 nm [2], d is dimension of Euclidean space, in which fractal is considered (it is obvious, that in our case $d=3$), d_{surf} is fractal dimension of nanofiller particles aggregate surface.

The value l_{st} is determined as follows [10]:

$$l_{st} = l_0 C_\infty, \qquad (2)$$

where l_0 is the main chain skeletal bond length, which is equal to 0.154 nm for both blocks of butadiene-styrene rubber [11], C_∞ is characteristic ratio, which is polymer chain statistical flexibility indicator [12] and is determined with the aid of the equation [9]:

$$T_g = 129\left(\frac{S}{C_\infty}\right)^{1/2}, \qquad (3)$$

where T_g is glass transition temperature, equal to 217 K for butadiene-styrene rubber [3], S is macromolecule cross-sectional area, determined for the mentioned rubber according to the additivity rule from the following considerations. As it is known [13], the macromolecule diameter quadrate values are equal: for polybutadiene – 20.7 Å^2 and for polystyrene – 69.8 Å^2. Having calculated cross-sectional area of macromolecule, simulated as a cylinder, for the indicated polymers according to the known geometrical formulas, let us obtain 16.2 and 54.8 Å^2, respectively. Further, accepting as S the average value of the adduced above areas, let us obtain for butadiene-styrene rubber $S=35.5$ Å^2. Then according to the equation (3) at the indicated values T_g and S let us obtain $C_\infty=12.5$ and according to the equation (2) – $l_{st}=1.932$ nm.

The fractal dimension of nanofiller surface d_{surf} was determined with the aid of the equation [3]:

$$S_u = 410R_p^{d_{surf}-d}, \tag{4}$$

where S_u is nanoshungite particles aggregate specific surface, calculated as follows [14]:

$$S_u = \frac{3}{\rho_n R_p}, \tag{5}$$

where ρ_n is nanofiller particles aggregate density, determined according to the formula [3]:

$$\rho_n = 0.188(2R_p)^{1/3}. \tag{6}$$

The calculation according to the equations (4)-(6) gives d_{surf}=2.44. Further, using calculated by the indicated mode parameters, let us obtain from the equation (1) the theoretical value of interfacial layer thickness l_{if}^T =7.8 nm. This value is close enough to the obtained one experimentally (their discrepancy makes up ~ 10 %).

The second method of value l_{if}^T estimation consists in using of the two following equations [1, 15]:

$$\varphi_{if} = (d_{surf} - 2)\varphi_n \tag{7}$$

and

$$\varphi_{if} = \varphi_n\left[\left(\frac{R_p + l_{if}^T}{R_p}\right)^3 - 1\right], \tag{8}$$

where φ_{if} and φ_n are relative volume fractions of interfacial regions and nanofiller, accordingly.

The combination of the indicated equations allows to receive the following formula for l_{if}^T calculation:

$$l_{if}^T = R_p\left[(d_n - 1)^{1/3} - 1\right]. \tag{9}$$

The calculation according to the formula (9) gives for the considered nanocomposite l_{if}^T =10.8 nm, that also corresponds well enough to the experiment (in this case discrepancy between l_{if} and l_{if}^T makes up ~ 19 %).

Let us note in conclusion the important experimental observation, which follows from the processed by programme SPIP results of the studied nanocomposite surface scan (Figure 1). As one can see, at one nanoshungite particle surface from one to three (in average – two) steps can be observed, structurally identified as interfacial layers. It is significant that these steps width (or l_{if}) is approximately equal to the first (the closest to nanoparticle surface) step width. Therefore, the indicated observation supposes, that in elastomeric nanocomposites at average two interfacial layers are formed – the first – at the expence of nanofiller particle surface with elastomeric matrix interaction, as a result of that molecular mobility in this layer is frozen and its state is glassy-like, and the second – at the expence of glassy interfacial layer with elastomeric polymer matrix interaction. The question, whether nanocomposite one interfacial layer or both serve as reinforcing element, is the most important from the practical point of view. Let us fulfil the following quantitative estimation for this question solution. The reinforcement degree (E_n/E_m) of polymer nanocomposites is given by the equation [3]:

$$\frac{E_n}{E_m} = 1 + 11\left(\varphi_n + \varphi_{if}\right)^{1.7},$$
(10)

where E_n and E_m are elasticity moduli of nanocomposite and matrix polymer, accordingly (E_m=1.82 MPa [3]).

According to the equation (7) the sum ($\varphi_n + \varphi_{if}$) is equal to:

$$\varphi_n + \varphi_{if} = \varphi_n\left(d_{surf} - 1\right),$$
(11)

if one interfacial layer (the closest to nanoshundite surface) is a reinforcing element and

$$\varphi_n + 2\varphi_{if} = \varphi_n\left(2d_{surf} - 3\right),$$
(12)

if both interfacial layers are a reinforcing element.

In its turn, the value φ_n is determined according to the equation [16]:

$$\varphi_n = \frac{W_n}{\rho_n},$$
(13)

where W_n is nanofiller mass content, ρ_n is its density, determined according to the formula (6).

The calculation according to the equations (11) and (12) gave the following E_n/E_m values: 4.60 and 6.65, accordingly. Since the experimental value E_n/E_m=6.10 is closer to the value, calculated according to the equation (12), then this means that both interfacial layers are reinforcing element for the studied nanocomposites. Therefore the coefficient 2 should be introduced in the equations for value l_{if} determination (for example, in the equation (1)) in case of nanocomposites with elastomeric matrix. Let us remind, that the equation (1) in its initial form was obtained as a relationship with proportionality sign, i.e. without fixed proportionality coefficient [8].

CONCLUSIONS

Nanoscopic methodics used in the present paper allow to estimate both interfacial layer structural special features in polymer nanocomposites and its sizes and properties. For the first time it has been shown, that in elastomeric particulate-filled nanocomposites two consecutive interfacial layers are formed, which are reinforcing element for the indicated nanocomposites. The proposed theoretical methodics of interfacial layer thickness estimation, elaborated within the frameworks of fractal analysis, give well enough correspondence to experiment.

REFERENCES

[1]   Kozlov G.V., Burya A.I., Lipatov Yu.S. *Mekhanika Kompozitnykh Materialov*, 2006, v. 42, № 6, p. 797-802.

[2]   Kornev Yu.V., Yumashev O.B., Zhogin V.A., Karnet Yu.N., Yanovskii Yu.G., Gamlitskii Yu.A. Kauchuk i Rezina, 2008, № 6, p. 18-23.

[3]   Mikitaev A.K., Kozlov G.V., Zaikov G.E. Polymer Nanocomposites: the Variety of Structural Forms and Applications. New York, Nova Science Publishers, Inc., 2008, 319 p.

[4]   Oliver W.C., Pharr G.M. *J. Mater. Res.,* 1992, v. 7, № 6, p. 1564-1583.

[5]   Lazorenko M.V., Badlyuk S.V., Rokochii N.V., Shut N.I. Kauchuk i Rezina, 1988, № 11, p. 17-19.

[6]   Kozlov G.V., Yanovskii Yu.G., Lipatov Yu.S. Mekhanika *Kompozitsionnykh Materialov i Konstruktsii,* 2002, v. 8, № 1, p. 111-149.

[7]   Kozlov G.V., Yanovskii Yu.G., Zaikov G.E. Structure and Properties of Particulate-Filled Polymer Composites: the Fractal Analysis. New York, Nova Science Publishers, Inc., 2010, 341 p.

[8]   Hentschel H.G.E., *Deutch J.M. Phys. Rev.* A, 1984, v. 29, № 3, p. 1609-1611.

[9]   Kozlov G.V., Zaikov G.E. Structure of the Polymer Amorphous State. Utrecht, Boston, Brill Academic Publishers, 2004, 465 p.

[10] Wu S. J. Polymer Sci.: Part B: Polymer Phys., 1989, v. 27, № 4, p. 723-741.

[11] Aharoni S.M. *Macromolecules,* 1983, v. 16, № 9, p. 1722-1728.

[12] Budtov V.P. Physical Chemistry of Polymer Solutions. Sankt-Peterburg, Khimiya, 1992, 384 p.

[13] Aharoni S.M. *Macromolecules*, 1985, v. 18, № 12, p. 2624-2630.

[14] Bobryshev A.N., Kozomazov V.N., Babin L.O., Solomatov V.I. Synergetics of Composite Materials. Lipetsk, NPO ORIUS, 1994, 154 p.

[15] Kozlov G.V., Yanovskii Yu.G., Karnet Yu.N. *Mekhanika Kompozitsionnykh Materialov i Konstruktsii*, 2005, v. 11, № 3, p. 446-450.

[16] Sheng N., Boyce M.C., Parks D.M., Rutledge G.C., Abes J.I., Cohen R.E. *Polymer,* 2004, v. 45, № 2, p. 487-506.

In: Polymer Yearbook – 2011.
Editors: G. Zaikov, C. Sirghie et al. pp. 175-182

ISBN 978-1-61209-645-2

Chapter 17

STRUCTURE FORMATION SYNERGETICS AND PROPERTIES OF NANOCOMPOSITES POLYPROPYLENE/CARBON NANOTUBES

G. V. Kozlov[1)]and G. E. Zaikov [2)]

[1)]Kabardino-Balkarian State University,[1]Nal'chik-360004, Russian Federation
[2)]N.M. Emanuel Institute of Biochemical Physics
of Russian Academy of Sciences,[2] Moscow-119334, Russian Federation

ABSTRACT

It has been shown, that aggregation (tangled coils formation) of carbon nanotubes begins at their very small contents. This factor strongly reduces reinforcement degree of nanocomposites polymer/carbon nanotubes. The estimation of the main parameters, influenced on the indicated nanocomposites elasticity modulus, was fulfilled. Theoretical calculations showed high potential of nanocomposites filled with nanotubes.

Keywords: Nanocomposite, carbon nanotubes, aggregation, reinforcement degree, interfacial adhesion, synergetics.

INTRODUCTION

Recent studies [1, 2] allow to distinguish the carbon nanotubes (CNT) as the perspective objects, allowing to create materials with principally new properties. The discovered in 90-s of the last century CNT possess unique characteristics: large strength in combination with elastic strain high values, good conductivity and adsorbable properties, ability to electrons cold emission and gases accumulation, chemical and thermal stability and so on.

[1] Chernyshevsky st., 173, Nal'chik-360004, Russian Federation.

[2] Kosygin st., 4, Moscow-119334, Russian Federation.

There is a large number of CNT modifications, differing in layers number, sizes, network structure form and, hence, in properties. High CNT cost for onelayered tubes (it makes up several tens and even hundreds of dollars USA per gram) dictates selection of those nanomaterials which together with inherent high qualitative proofs are accessible to industrial producers of nanocomposites both from the point of view of production amounts and in the plan of commercial perspectives of the product realization.

As it was shown earlier [3], polymer nanocomposites filled with CNT had a number of specific features. Therefore the purpose of the present paper is the study of structure and reinforcement degree of nanocomposites polypropylene/CNT (PP/CNT) [4] with accounting for the indicated above specific features.

EXPERIMENTAL

Polypropylene of the industrial production mark Kaplen 01030 was used as matrix polymer and multilayered CNT, having specific surface of 130-150 m^2/g, layers number of 20-30 and external diameter of 20-30 nm were used as nanofiller. The contents of CNT was varied within the frameworks of 0.15-3.0 mass %.

Nanocomposites PP/CNT were prepared by components mixing in melt on twin-screw extruder Thermo Haake model Reomex RTW 5567, production of German Federal Republic. Mixing was performed at temperature 463-503 K and screw speed of 150 rpm during 5 min. Specimens for tension tests in the form of two-sided spade with sizes according to GOST-12423-66 were produced by casting under pressure method on casting machine Test Samples Molding Apparate RR/TS MP of firm Ray-Ran (Taiwan) at temperature 503 K and pressure 8 MPa [4].

Mechanical tests on uniaxial tension were carried out on universal testing apparatus Gotech Testing Machine CT-TCS, production of German Federal Republic, at temperature 293 K and strain rate $\sim 2 \times 10^{-3}$ s^{-1}.

RESULTS AND DISCUSSION

To estimate nanofiller (CNT) volume contents φ_n is possible according to the known equation [5]:

$$\varphi_n = \frac{W_n}{\rho_n},\qquad (1)$$

where W_n is CNT mass contents, ρ_n is CNT density, estimated according to the equation [3]:

$$\rho_n = 0.188(D_{CNT})^{1/3},\qquad (2)$$

where D_{CNT} is nanotubes diameter.

In Figure 1 the experimentally received dependence of reinforcement degree E_n/E_m (E_n and E_m are elasticity moduli of nanocomposite and matrix polymer, accordingly) on volume filling degree φ_n. As one can see, at very small values φ_n of order of 0.015 the indicated dependence reaches plateau with small E_n/E_m values about 1.20. Let us note, that such magnitudes of reinforcement degree at $\varphi_n \approx 0.06$ are typical for microcomposites, i.e. polymer composites with filler of micron sizes [6]. Let us consider the reasons of nanocomposites PP/CNT such behaviour.

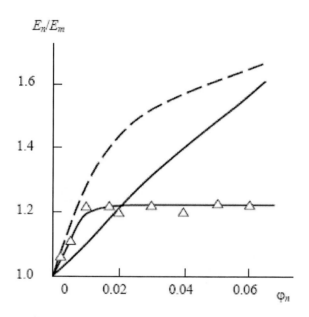

Figure 1.The dependences of reinforcement degree En/Em on nanofiller volume contents φn. 1, 2 – the experimental data for nanocomposites PP/CNT (1) and LDPE/CaCO3 (2), 3 – theoretical calculation according to the equation (5).

Within the frameworks of percolation model the value E_n/E_m estimation can be performed by using of the following relationship [3]:

$$\frac{E_n}{E_m} = 1 + 11\left(\varphi_n + \varphi_{if}\right)^{1.7},\qquad(3)$$

where φ_{if} is interfacial regions relative fraction, which in CNT case is connected with value φ_n as follows [3]:

$$\varphi_{if} = 2.86\varphi_n b,\qquad(4)$$

where b is parameter, characterizing interfacial adhesion level in polymer composites [7].

The parameter b allows to clear gradation of interfacial adhesion level. So, the condition $b=0$ means interfacial adhesion absence, $b=1.0$ – perfect (by Kerner) adhesion and the condition $b>1.0$ defines nanoadhesion effect [8]. Let us note, that for polymer microcomposites with different fillers and matrix polymers b variation makes up $\sim$ -0.19-1.39 [7].

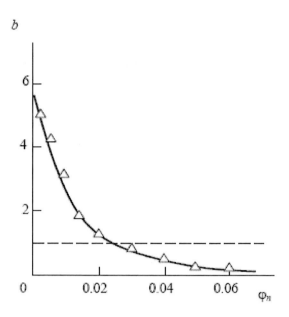

Figure 2.The dependence of parameter b on CNT volume contents φn for nanocomposites PP/CNT. The horizontal shaded line shows perfect adhesion level (b=1).

The parameter b calculation according to the equations (3) and (4) shows its reduction from 5.09 up to 0.21 in the range of φ_n=0.003-0.060 (Figure 2). As it follows from the data of Figure 2, CNT contents increase in the studied nanocomposites results to interfacial adhesion level qualitative changes: at $\varphi_n \leq 0.020$ the nanoadhesion effect is observed, at $\varphi_n \approx 0.025$ the perfect (by Kerner) adhesion is realized at $\varphi_n > 0.030$ b→0 as well.

The authors [9] considered three main cases of the dependence of reinforcement degree E_n/E_m on nanofiller volume contents φ_n. Although the indicated treatment was obtained for particulate-filled polymer composites, its application to the studied nanocomposites is of a definite interest. There are the following main types of the dependences $E_n/E_m(\varphi_n)$:

1) perfect adhesion between nanofiller and polymer matrix, described by Kerner's equation, which can be approximated by the following relationship:

$$\frac{E_n}{E_m} = 1 + 11.6\varphi_n - 44.4\varphi_n^2 + 96.3\varphi_n^3;$$ (5)

2) zero adhesional strength at large friction coefficient between nanofiller and polymer matrix, which is described by the equation:

$$\frac{E_n}{E_m} = 1 + \varphi_n;$$ (6)

3) the interaction complete absence and ideal sliding between nanofiller and polymer matrix, when composite elasticity modulus is practically determined by polymer cross-section and is connected with filling degree by the equation:

$$\frac{E_n}{E_m} = 1 - \varphi_n^{2/3}.$$ (7)

In Figure 1 theoretical dependence $E_n/E_m(\varphi_n)$, calculated according to the equation (5) (curve 3), is adduced. Its comparison with the experimentally obtained dependence $E_n/E_m(\varphi_n)$ (curve 1) shows, that at $\varphi_n=0.003-0.015$ (the nanoadhesion effect realization) the experimental values E_n/E_m exceed theoretical reinforcement degree, at $\varphi_n\approx0.020$ these magnitudes are equal (as it follows from the data of Figure 2, at the indicated filling degree $b=1.0$, i.e. perfect adhesion is realized) and at $\varphi_n>0.020$ theoretical reinforcement degree is higher than the experimental one, that is due to further reduction of interfacial adhesion level at φ_n growth (Figure 2).

Let us consider the reasons of sharp reduction of interfacial adhesion level, characterized by parameter b, at CNT volume contents growth. CNT orientation factor η is determined as follows [10]:

$$\varphi_{if} = 1.09\eta,$$ (8)

where interfacial regions relative fraction φ_{if} can be calculated with the aid of the equation (3).

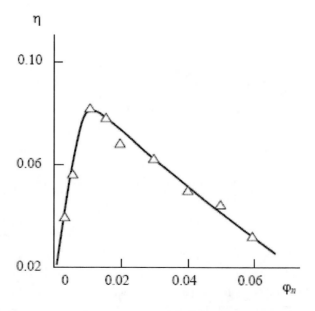

Figure 3.The dependence of CNT orientation factor η on their volume contents φn for nanocomposites PP/CNT.

In Figure 3 the dependence of orientation factor η on filling volume degree φ_n for nanocomposites PP/CNT is adduced, which has pronounced extreme character with maximum at φ_n=0.010-0.015. Such character of the dependence $\eta(\varphi_n)$ assumes (together with similar behaviour of other properties number for the studied nanocomposites – yield stress σ_Y, impact toughness A_p etc. [4]) the common feature, having statistical character: at first periodic (ordered) behaviour is observed, which is close to a sigmoid one with period doubling and then the transition to chaotic behaviour is realized. Such behaviour is typical for synergetic systems [11] and it was observed for nanocomposites phenilone/CNT [12]. However, the latter were processed in rotating electromagnetic field that reduces essentially nanotubes aggregation degree. Quantitatively this effect is expressed in maximum displacement from φ_n=0.015 for nanocomposites PP/CNT up to φ_n=0.064 for nanocomposites phenylone/CNT [3, 12]. In addition, reinforcement degree for the first from the indicated nanocomposites is less than for the second ones: E_n/E_m=1.223 and 1.416, accordingly. It is significant, that the second value is reached simply at the expence of φ_n larger value at practically equal η magnitudes. Hence, CNT periodic behaviour consists in their partial orientation in polymer melt and chaotic behaviour means CNT tangled coils formation with final value η=0 [12]. Therefore, the stated above results suppose, that the condition $E_n/E_m \approx 1.22$=const for nanocomposites PP/CNT at $\varphi_n \geq 0.015$ (Figure 1) is defined by φ_n increase compensation at the expence of φ_{if} reduction (the equation (3)), due to CNT aggregation or η decreasing (the equation (8)). In its turn, φ_{if} decreasing is due to reduction of interfacial adhesion level, characterized by the parameter b, in virtue of the same CNT aggregation.

It is often supposed [13] that at present CNT are the most perspective nanofiller for polymer nanocomposites production. Similar opinion was repeatedly stated earlier in respect of organoclay [14]. In Figure 1 the dependence $E_n/E_m(\varphi_n)$ is adduced, where φ_n is calculated according to the equations (1) and (2), for particulate-filled nanocomposites low density polyethylene/calcium carbonate (LDPE/CaCO$_3$) [15]. As it follows from the dependences $E_n/E_m(\varphi_n)$ comparison for nanocomposites PP/CNT and LDPE/CaCO$_3$, the advantage of the latter is obvious. Thus, it is impossible to assert, that the nanofiller definite type has some principal advantages. Polymer nanocomposites final properties are defined by parameters set: interfacial adhesion level, nanofiller aggregation degree, polymer matrix correct choice and so on [3]. But there is another aspect of the problem both technological and economic. From this point of view at present disperse particles are the most attractive, which are cheap, easily processed by a binding agent, and simply dispersed. At the same time it's unlikely to obtain at present exfoliated organoclay at W_n>3 mass %. Still it's more difficult to obtain separate CNT, but not in the form of tangled coils, which are in addition very expensive, as it was indicated above. Therefore nanocomposites polymer/CNT application as engineering materials will unlikely be practiced on a wide scale in the very near future, that does not exlude their usage in specific applications [2, 14, 16]. Nevertheless, theoretical perspectiveness of CNT application in the indicated capacity is also obvious. So, at the paper obtained in present (and, hence, real ones) the largest values η=0.083, b=5.09 and at φ_n=0.060 (W_n=3 mass %) nanocomposite PP/CNT elasticity modulus can reach ~ 9210 MPa. However, the same elasticity modulus can be obtained for nanocomposites on the basis of PP at the expence of condition φ_{if}=1.225$\varphi_n b$ [3] and b=5.09 at disperse nanofiiler with diameter of 20 nm contents 6.3 mass % only or with diameter of 50 nm – 8.0 mass %. It is obvious, that technologically and economically the last variant is more preferable.

And let us note in conclusion one more important methodological aspect. As distinct from widely used micromechanical models [5, 6], not a single one from the adduced above equations does not used nanofiller elasticity modulus as parameter. Such approach is typical for percolation [17] and fractal [18] polymer composites (nanocomposites) reinforcement models. The indicated treatment is confirmed in practice: CNT exceptional high elasticity modulus [14] does not give them an advantage in comparison with organoclay or disperse nanofiller (Figure 1).

CONCLUSIONS

The obtained in the present paper results show, that carbon nanotubes aggregation (tangled coils formation) begins with the last very small contents and restricts strongly the obtained nanocomposites elasticity modulus enhancement. The offered model does not take into consideration nanofiller elasticity modulus and shows, that the main factors in polymer nanocomposites in general properties are interfacial adhesion (nanoadhesion) level, nanofiller aggregation degree, polymer matrix choice. The performed theoretical estimations show high potential of polymer nanocomposites.

REFERENCES

[1]   Pool C., Owens F. Nanotechnologies. Moscow, Technosphere, 2005, 336 p.
[2]   Rakov E.G. Uspekhi Khimii, 2007, v. 76, № 1, p. 3-19.
[3]   Mikitaev A.K., Kozlov G.V., Zaikov G.E. Polymer Nanocomposites: the Variety of Structural Forms and Applications. New York, Nova Science Publishers, Inc., 2008, 319 p.
[4]   Chukov N.A., Molokanov G.O., Dzhangurazov B.Zh., Danilova-Volkovskaya G.M., Khashirova S.Yu., Mikitaev A.K. Proceedings of II Internat. Sci.-Part. Conf. "Nanostructures in Polymers and Polymer Nanocomposites". Nal'chik, KBSU, 2009, p. 147-150.
[5]   Sheng N., Boyce M.C., Parks D.M., Rutledge G.C., Abes J.I., Cohen R.E. Polymer, 2004, v. 45, № 2, p. 487-506.
[6]   Ahmed S., Jones F.R. *J. Mater. Sci.,* 1990, v. 25, № 12, p. 4933-4942.
[7]   Holliday L., Robinson J.D. In book: Polymer Engineering Composites. Ed. Richardson M.O.W. London, Applied Science Publishers LTD, 1978, p. 241-283.
[8]   Kozlov G.V., Burya A.I., Yanovskii Yu.G., Zaikov G.E. *J. Balkan Tribologic. Assoc.,* 2008, v. 14, № 2, p. 171-177.
[9]   Tugov I.I., Shaulov A.Yu. *Vysokomolek. Soed.* B, 1990, v. 32, № 7, p. 527-529.
[10] Kozlov G.V., Yanovskii Yu.G., Zaikov G.E. Synergetics and Fractal Analysis of the Polymer Composites Filled with Short Fibers. New York, Nova Science Publishers, Inc., 2010 (in press).
[11] Ivanova V.S., Kuzeev I.R., Zakirnichnaya M.M. Synergetics and Fractals. Universality of Materials Mechanical Behaviour. Ufa, Publishers USNTU, 1998, 366 p.

[12] Kozlov G.V., Burya A.I., Zaikov G.E. J. Balkan Tribologic. Assoc., 2007, v. 13, № 4, p. 475-479.

[13] Burya A.I., Tkachev A.G., Mitschenko S.V., Nakonechnaya N.I. Plast. Massy, 2007, № 12, p. 36-41.

[14] Eletskii A.V. Uspekhi Fizicheskikh Nauk, 2007, v. 177, № 3, p. 223-274.

[15] Syltonov N.Zh., Mikitaev A.K. Proceedings of VI Internal. Sci.-Pract. Conf. "New Polymer Composite Materials". Nal'chik, KBSU, 2010, p. 392-398.

[16] Buchachenko A.L. Uspekhi Khimii, 2003, v. 72, № 5, p. 419-437.

[17] Bobryshev A.N., Kozomazov V.N., Babin L.O., Solomatov V.I. Synergetics of Composite Materials. Lipetsk, NPO ORIUS, 1994, 154 p.

[18] Kozlov G.V., Yanovskii Yu.G., Zaikov G.E. Structure and Properties of Particulate-Filled Polymer Composites: the Fractal Analysis. New York, Nova Science Publishers, Inc., 2010, 341 p.

In: Polymer Yearbook – 2011.
Editors: G. Zaikov, C. Sirghie et al. pp. 183-194

ISBN 978-1-61209-645-2
© 2011 Nova Science Publishers, Inc.

Chapter 18

GRADIENTLY ORIENTED STATE OF POLYMERS: FORMATION AND INVESTIGATION

Levan Nadareishvili[1], Zurab Wardosanidze[1], Nodar Lekishvili[2], Nona Topuridze[1], Gennady Zaikov[3] and Ryszard Kozlowski[4]

[1]Cybernetics Institute[1], 0186 Tbilisi, Georgia
[2]Ivane Javakhishvili Tbilisi State University, Faculty of Exact and Natural Sciences[2]
0128 Tbilisi, Georgia
[3]N.M. Emanuel Institute of Bichemical Physics RAS[3], 119991 Moscow, Russia
[4]FAO ESCORENA, [4] Poznan 60-630, Poland

ABSTRACT

For the first time there is formulated a notion of gradiently oriented state. Some regularity for formation of gradiently oriented state in polymers are established .The specificity of influence of non-homogeneous mechanical field has been demonstrated on the example of creation of GB-elements, which is the object of research of the new non-traditional direction of the gradient optics – GB-optics (Gradient Birefringence Optics). Possible spheres of use of GB-elements are polarized compensators, polarized holography and photonics, the interference polarized GB-monochromator, the luminescence analysis, etc.

Keywords: polymers, gradiently oriented state, regularity, GB-elements, GB-monochromator, luminescence analysis.

[1] 5, S. Euli, str., 0186 Tbilisi, Georgia, E-mail: levannadarei@yahoo.com.
[2] 1, Ilia Chavchavadze Ave., 0128 Tbilisi, Georgia; e-mail: nodar@lekishvili.info
[3] 4, Kosigin street, 119991 Moscow, Russia. E-mail: chembio@sky.chph.ras.ru
[4] 71B, Wojska Polskiego str., Poznan 60-630, Poland. E-mail: Ryszard.Kozlowski@escorena.net.

INTRODUCTION

Creation and investigation of materials with the gradient of properties is considered one of the main directions of polymer science for the 21st century [1].

In this direction, the essential success was achieved in the 70s of the previous century. The methods of obtaining materials with inhomogeneous distributions of properties were developed. The heterogeneity of composition conditions the gradient of refraction. In such an area a trajectory of the light beam is constantly curvilinear along the free path length, which causes deviation of light beam, and during proper distribution of refraction index – focusing of beams, too. In the 70s, the cylinder elements were fabricated from polymers with radial-parabolic distribution of composition, having the ability of transmission of light and focusing. These elements received the name Selfoc (Self focusing) [2-7].

By introducing a new parameter in optics - refraction index gradient, - new conditions of theoretical and experimental investigations have been created. The fundament was laid both for wide possibilities for creation of radical improvement of existing optical devices and new optical devices, fabrication of which on the basis of traditional optical materials have been excluded principally. In scientific literature for indication of refraction gradient elements (materials, areas) there has been established a term GRIN (Gradient Refractive Index), and for corresponding field of science – GRIN-optics.

For today GRIN–optics is the independent perspective direction. GRIN elements can independently form and translate an image without additional means. They are widely used in optics and opto-electronics of various destinations. (Fiber-optical lines of communications, the agreement elements, endoscopies systems of small sizes, copier, lenses of low chromatic aberrations, the focusing elements of video-recording laser systems, flat lenses, etc). In each technologically developed country the intensive investigations in this field are being carried out. The special laboratories and research centers have been created. We have received some interesting results in this sphere [8, 10-11].

The world practice of development of GRIN-optics indicates that the attention of the researchers is concentrated only on the refraction index [2-5].

At the same time, experts of GRIN optics consider similarly perspective the materials having other optical properties, in particular, materials of gradient birefringence, too [1].

The first polymer materials (films) having such optical properties were obtained by us [11]. We also contributed the acronym for the materials with gradient birefringence (elements, areas) GB (Gradient Birefringence) – material (element, area).

By introducing a new parameter in optics – Gradient Birefringence - a fundament of a new direction of optics was laid; it resulted in widening the notion of "Gradient Optics" [12-17]. Nowadays, the gradient optics covers two independent directions – GRIN–optics and GB-optics. Both these directions are generally strategic in development of polymeric science of gradient material science.

The ways of formation of GRIN and GB elements are different. The GRIN-elements are received as a result of chemical transformations. Many methods have been elaborated [9-11], by means of which the known chemical transformations are realized in monomer (polymeric) systems in gradient regime (in selected directions and proper velocity).

Formation of GB elements is based on ability of great deformations characteristic to polymers. At the same time while realizing great deformations polymer passes on to specific, so called oriented state. Its essence is expressed in the macromolecule chains (as usual on separate sections) of polymeric body that have privileges locations in the whole polymeric body in any direction which is called an orientation axis.

The most widely spread of polymeric orientation is mono-axial orientation stretching. As a result of the orientation the physical and chemical properties of the whole number of exploitation properties of polymers are changed essentially. That is why the studying of orientation processes which is an integral part of polymers structure study (molecular, super molecular) is one of the main tasks of polymer science.

With allowance of property gradient, as a strategic mark of development of polymeric science, the oriented state of polymers acquires a new essence, which can be qualified as gradiently oriented state.

We suppose that introduction of new structural characterization of polymers as a new physical characteristic of polymeric nature essentially broadens general problems of scientific research of polymers. It principally increases possibilities of regulation of mechanical, thermal, electric, optical and other properties, giving an impetus to creation of new scientific directions as it has already happened on the example of GB-optics [11].

The essential characteristic of this position except gradient is the angle between structural orientation and the direction of the gradient. We can regulate it in the interval $90^0 \geq \alpha \geq 0^0$.

For creation of gradiently oriented state the form of isotropic polymer sample (i.e., the form of clamps and inter-location) has been selected so that it is provided with preliminarily established gradient of relative lengthening in the sample. By means of variation of other parameters (temperature, value, velocity of deformations, etc.) it is possible to regulate gradiently oriented state of polymers. Characterization of this state is affected by means of observation on polarized light through the polarized microscope and studying of birefringence. Thus, the birefringence is a testing characteristic of gradiently oriented state and at the same time it is itself a purposeful property of the material.

Above the glazing temperature during the mono-axial linear polymer the polymer acquires symmetry of mono-axial crystal, the optical axis of which coincides with the stretching direction. The birefringence originated this time is functions of relative lengthening:

$$\Delta n = n_1 - n_2 = \gamma\lambda \qquad , \tag{1}$$

There: n_1 – is an ordinary ray; n_2 – extraordinary ray; γ - optical coefficient of deformation; λ - relative lengthen. The equation (1) is valid during the identity of all other parameters determined by the process mode.

We have elaborated out several versions of the equipment [17] for creation of preliminarily established inhomogeneous mechanical field, as a result of cohesion in a polymeric body an established gradient of relative lengthening is formed, and consequently, so is the preliminarily established gradient birefringence.

For one series of the equipment, **GB-** effect is achieved by the fact that the clamps allocated on one plane, in which a polymer film/plate of trapezoid shape rotates in inter-opposite direction around the parallel axis, the rims (edges) of clamps create φ angle at the rotation axis. In this case the distribution of lengthens (Δl) on the polymeric sample length (h) is expressed by the equation:

$$\Delta l = 2x.tg\varphi(1-\cos\alpha) \ ,$$
(2)

There: x is the length of the sample in a given point ($x = 0; x = h$); α is the angle of rotation of clamps.

Some of our results are published in ref.[11-18].

In this work we consider some regularities of formation of the gradiently oriented state, which is connected with some properties of constructive determinations of corresponding apparatus. We investigate some optical properties of gradiently oriented polymers. There are discussed possible spheres of their application.

EXPERIMENTAL

Experiment was carried out on PVS films (thickness is 80-100mk).Tension of the film was made on device apparatus of specific construction (T = 358K, velocity of tension = 20 mm/min.). For creation of non-homogeneous mechanical field we use clamps, having various configurations.

We studied optical properties (gradient of anisotropy) on polarization micro-photometer.

DERIVATION OF THE BASIS RELATIONSHIPS

Distribution of the relative lengthening is determined by profile of clamps $f_1(x)$ and $f_2(x)$, where independent variable x changes in the $[x_1, x_2]$ interval. $f_1(x)$ undergoes parallel displacing across the OY axis and $f_2(x)$ is fixed function. Let us designate the relative lengthening across the OY axis as $\Phi(x)$ and the value of parallel transfer as K , then, for the distribution of the relative displacing we have:

$$\Phi(x) = \frac{K}{f_1(x) - f_2(x)}$$
(3)

We can choose $\Phi(x)$ from the different class of function (linear, parabolic, hyperbolic, sinusoidal, etc.), take into account isochromic aspect of GB-element. From the equation (3) we have:

$f_1(x) = K/\Phi(x) + f_2(x)$ (4)

Let us consider different cases of $\Phi(x)$ function:

1) Linear function $\Phi(x) = ax+b$, then

$$f_1(x) = K/(ax+b) + f_2(x) \tag{5}$$

when a=1, b=0 and the profile of the first clamp is linear, i.e. $f_1(x) = c$, then

$$f_1(x) = K/x + C \tag{6}$$

Thus, the profile of the second clamp is hyperbola.

2) Parabolic function $\Phi(x) = ax2 + bx + c$, then:

$$f_1(x) = K/(ax^2 + bx + c) + C \tag{7}$$

When a =1, b = 0, c = 0,

$$f_1(x) = K/x^2 + C \tag{8}$$

Thus, the profile of the second clamp is also hyperbola.

3) Hyperbolic function $\Phi(x) = a/x$, then:

$$f_1(x) = Kx/a + C \tag{9}$$

In this case the form of the second clamp is linear.

4) Sinusoidal function $\Phi(x) = \sin x$, then:

$$f_1(x) = K/\sin x + C \tag{10}$$

In this case the form of the second form is so complicated, that its realization is available only in the definitely, strictly limited conditions.

Let us assume that the initial length of the sample in the x point is more than distance between two clamps, i.e. $\Delta x > [f_1(x) - f_2(x)]$

Then for the relative lengthen we have:

$$\Phi(x) = K/(f_1(x) - f_2(x) + \Delta x) \tag{11}$$

In this case the profiles of the clamps changes considerably:

a) when $\Phi(x) = ax + b$ (linear function), then

$$f_1(x) = K/(ax+b) + f_2(x) - \Delta x \tag{12}$$

Value of Δx is preliminarily chosen and always is linear function [i. e., $\Delta(x) = \alpha x + \beta$].
When $f_1(x) = C$, then

$$f_1(x) = K/(ax+b) + C - (\alpha x + \beta) = K/(ax+b) - \alpha x + \gamma \tag{13}$$

Where $\gamma = C - \beta$; when a = 1, b = 0, then:

$$f_1(x) = K/(x - \alpha x) + \gamma \tag{14}$$

So, the profile of the second clamp is the linear combination of the hyperbola and line.

b) When $\Phi(x) = ax2 + bx + c$ (parabolic function), then:

$$f_1(x) = K/(ax^2+bx+c) + C - (\alpha x + \beta) \tag{15}$$

In this case the profile of the second clamp is complicated function too.

c) When $\Phi(x) = a/x$ (hyperbolic function), then:

$$f_1(x) = K/(a/x + c) - (\alpha x + \beta) = (K/a)x + c - \alpha x - \beta \tag{16}$$

So, the profile of the second clamp in this case is linear.

The obtained results give us a chance to improve possibilities of existing optical devices. At the same time it will be possible to project and create optical devices of new generation.

The profile of the clamps with all those factors, which have influence on the oriented state of polymers (molecular and super- molecular structure, medium molecular mass and distribution of molecular mass, existence of ingredients and plastificators, velocity of tensile deformation, values of relative deformation, temperature, environment, scale factor) define optical properties of GB-elements.

We represent the experimental results to illustrate some specific cases of formation of gradiently oriented state (creation of GB-elements).

PARALLEL CLAMPS

In this case the value of the relative displacement along of all the film perpendicularly to the direction of tension is constant. In accord to this there will be no gradient of birefringence. However, in polaroscope (crossed nicols, 6-times tension; the initial length of the film is 6 cm) sharp isochromes are seen on free ends of the film on the width of 1 cm. So, at the edge of the film the gradient of birefringence is in direction, perpendicular to the tension. There is a homogeneous region in the middle of the sample, where the birefringence (Δn) is constant. In Figure1a we have the microphotogram of film for the wavelength λ =630 nm that is obtained by means of photoelectric polarization microphotometer (crossed nicols). Figure.1a corresponds to the picture observed in polariscope. There was shown that in free region of given wavelength two isochromic bands are received, maximum of which correspond to the half-wave regions $(2n+1)\lambda/2$. Consequently, theoretically, a gradient will not take place when tension is parallel, but experimentally we observe that the gradient is received. The reason of this may be free edges of the film. Absolute lengthening at the edges is greater, than in the middle. Such effect takes place in all cases of tension.

CLAMPS DIRECTED BY 45^0 ANGLES

In this case $f_1(x) = (\sqrt{2}/2)x$ and $f_2(x)=(\sqrt{2}/2)x$. Then the equation is transformed into:

$$\Phi(x) = K/x\sqrt{2} \qquad\qquad (17)$$

This is an equation of hyperbola. When we use such clamps for parallel tension of the film, we receive an oriented polymer film, where regulation of distribution of birefringence gradient perpendicularly to the tension direction is hyperbolic. Indeed, the regulation of distribution in perpendicular to the tension direction is hyperbolic. In Figure1,b we give microphotograms of these samples. Here the distances between bands corresponding to $(2n+1)\lambda/2$ are nearly the hyperbolic distribution.

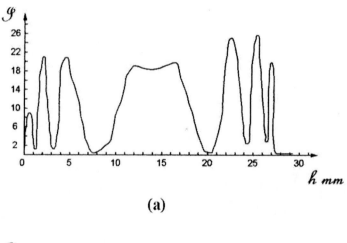

(a)

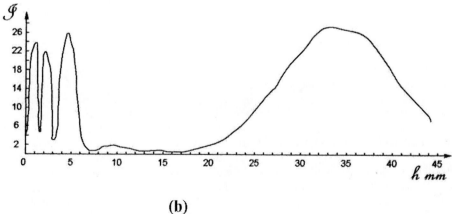

(b)

Figure 1. (Continued on next page)

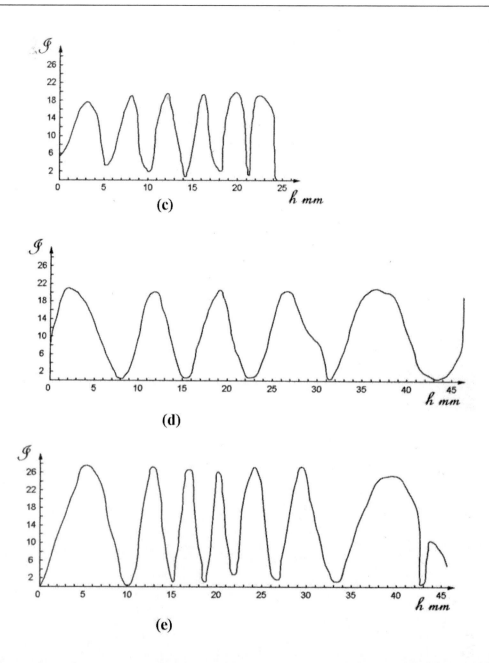

Figure 1. Dependence of transparency I of gradiently oriented film on h-coordinate (crossed Nichols) a) Parallel clamps, the film of rectangle form, tension is perpendicular to the clamps.b) Clamps allocated to the angle of 450 towards each other. Trapezium form Polymer film. The tension occurs in parallel to the base of trapezium. c and d) Parallel clamps. Trapezium form Polymer film. Direction of tension is parallel to the attitude of trapezium. The length of one of free edges of the film (l2) is more than the distance between clamps l1. .e) Parallel clamps. Trapezium form Polymer film. Direction of tension is parallel to the attitude of trapezium. Relation l2 / the distance between two clamps are more for one film (case c) than that for another (case d).3. Parallel clamps. The length of one of free edges of the film is more than a distance between two clamps (Figure 2).Parallel clamps. The length of one of free edges of the film is more than a distance between two clamps (Figure 2).

In this case, according to equation (9), the distribution of the relative lengthens may be various. The tension is carried out until distance between two clamps equals to l_2. In Figures 1,c and 1,d we have micro photogram and we see that the distances between bands corresponding to $(2n + 1) \lambda /2$ are practically equal.

At the same time relation l_2 / the distance between two clamps is more for one film (Figure1, c) than that for another (Figure1, d).

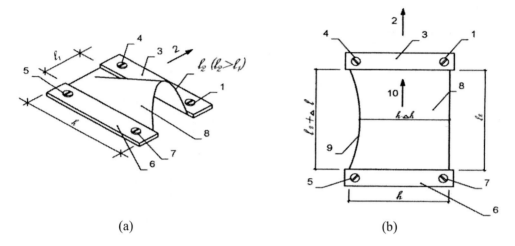

(a)   (b)

Figure 2. Gradient tension of polymer film 1,4,5,7 - screws; 2 – direction of tension; 3,6 - clamps; 8 - polymer film: a - before tension; b - after tension.

PARALLEL CLAMPS. ISOSCELES TRAPEZIUM, TENSION ACROSS THE ATTITUDE OF TRAPEZIUM

In this case isochromes are disposed near the big base of trapezium and parallel to it. In Figure1e there is given corresponding microphotogram. In all cases described previously (cases 1, 2, 3) direction of tension, disposition of isochromes and gradient to each other is described by the scheme on Figure3a. This case is in contrast to other ones. Here disposition of these characteristics is different and is described by the scheme on Figure3b.

Polarized Compensators: Generally, the main principle of behavior of compensators is based on operation of exclusion of any optical parameter. For example, in case of ordinary isotropic phase compensators the thickness and refraction index ($n \cdot d$) of optical element have to provide with concrete phase shifting ($k.\lambda,...\infty$). Such compensators are used in interferometer tasks. In case of anisotropy compensators the notion of phase shift means not the absolute phase shifting for a given wavelength, but rather the relative phase shifting $\Delta n \cdot d$) between usual and unusual beams (waves).That is why that the accuracy of optical measurements have been increased considerably by using the anisotropy compensators. In spite of the aforementioned, the area of practical use of both the isotropic and anisotropy compensators is limited, so as practically they fulfill the function of a reference.

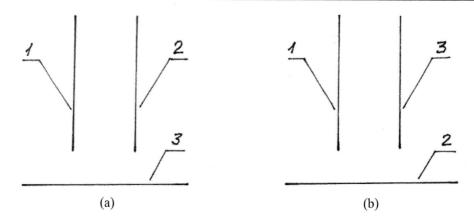

Figure 3. Direction of tension, disposition of isochromes and gradient in gradiently oriented films: 1.-
Direction of tension; 2. - Disposition of isochromes; 3.- Gradient: a) Cases 1, 2, 3; b) Case 4.4. Some
spheres of application of gb-elements

In comparison to them, the GB compensator is not a reference optical element but it
provides with both the concrete phase shifts in the whole section of visible spectrum and any
phase shift for the given wavelength.

Polarized Holography and Photonics: Both in holography and photonics it has become
necessary to envisage the polarized characteristics of the light. Generally in this case it is
important only the intensity gradation was the reason why dynamic range of the process was
sharply limited. In case of fixation of polarization it is already decisive not the intensity
gradation (which is limited), but the fixation of practically boundless (infinite) versions of
polarization. As a consequence in a given case it becomes necessary to provide with complex
space polarized modulation of light. One of the ways for solution of this task is to use the GB
elements having the complicated structure which gives possibility to realize the formation of
light waves having the specific polarized characteristics.

A special importance is given to applying GB elements in photonics (in photochemistry),
so as for today the investigations are being carried out very intensively for studying both the
liner and nonlinear Veigert effects [19]. This time the GB elements will provide with the
accurate and simple relation between the linear and nonlinear Veigert effects. In particular,
the calibrated (divided) GB elements give possibility to realize the radiation simultaneously
by means of all kinds of polarization. This time it is very important that in corresponding
sections of polarization the power exposition is absolutely similar, which is automatically
realized. And this is particularly important, since it excludes the necessity of labor-intensive
and less accurate photometric measurements. In photonics the GB elements will fulfill the
function a definite standard polarized modulators.

The Interference Polarized GB-Monochromator: the issue of miniaturization of spectral
devices is very actually. It is clear that there are already created miniature monochromators in
the form of gradient multi-layer filters. The indicated filters are distinguished by the fact that
the thickness of separate dielectric layers of multi-layer systems is constantly changed along
the filter which ensures a maximum transparency in a red section of spectrum on one its part,
and in the violet section on the other side [20]. The lack of such monochromators is the low
capability of spectral distinction ($>10A^{0}$). It is possible to improve the capability of spectral

distinction if the interference-polarization filter is created under the similar principles. At the same time in such a monochromator it will be used not only the multi-layer structure of alternative thickness, but the additional gradient anisotropic structure (GB-structure), the multiple of the wave-length of which will be agreed-compared ($\Delta n \cdot d = n \cdot \lambda$) with the maxims of transparency spectrum of the monochromator. The capability of distinguishing of such a monochromator is of 0.1 A^0 order. The sizes of a monochromator may be from 5.1-mm^2 to 50-100 mm^2.

The Luminescence Analysis: It seems very interesting the application of anisotropic films in luminescence analysis. In particular, as it is known in oriented organic films as in matrix. Radiation of installed luminescence paints is partially polarized. Besides, this time the degree luminescence (fluorescence) polarization mainly depends on the structure of luminescence material molecule itself and the quality of matrix orientation. In case if the matrix orientation changes along the film, i.e. we have a gradient anisotropic matrix, the luminescence polarization quality is characterized too by definite distribution, i.e. by the gradient to the same direction. If the degree of luminescence polarization, let's say along the X axis, is expressed by $P_1(x)$, and the degree of matrix polarization or distribution of anisotropy - by $P_m(x)$ or $\Delta n(x)$, than the relative value:

$$K = P_1(x) / P_m(x) \text{ or } K = P_1(x) / \Delta n(x) \tag{18}$$

in totally with luminescence intensity and spectral characteristics unambiguously should determine the definite features of given luminescence materials. This approach has definite properties as well for possible broadening of investigation of laser effects.

For development of GB-optics (similarly as of classical optics) processing/development is a universal method which allows us to realize GB-structures of concrete functional destination by means of creation of various technological equipment and controlling of their technical parameters. In case of simple one-dimensional compensators the main determining technological parameters can be the relative lengthens, profile and scale of clamps. In case of relatively complex two-dimensional compensators to this three main parameters are added the coordinated parameters which mean the simultaneous orientation to the orthogonal or generally to any other direction, too. If we add to these materials a space modulation of temperature field as well, e.g. gradient heating the obtained GB element configuration can be changed within the wide ranges according to their destination.

BASIC CONCLUSIONS

- For the first time there was formulated a notion of gradient oriented state.
- Some regularities of formation of gradiently oriented state of polymers were established.
- .Some concrete cases of creation GB-elements were discussed.
- .Possible spheres of using GB-elements were discussed.

REFERENCES

[1]   Lekishvili N., Nadareishvili L., Zaikov G., Khananashvili L. Polymers and polymeric materials for fiber and gradient optics. VSP (Utrecht-Boston-Köln-Tokyo), 2002.

[2]   Chen. W.C, ChenI H., Yang S.Y. et. al. Photonic and Optoelectronic Polymers.

[3]   By Ed. Samson A. Lenkhe, Keneth E.Wynnei. Chapter 6. A.C.S. Washington, DC, 1997.

[4]   Sheng Pin Wu, YEisuke N., Koike Y. *Appl. Opt.* 1996, 35,1, 28.

[5]   Shvartsburg A. and Petite G. Optics Letters 2006, 31, 1127.

[6]   Tagaya A., Koike Y. Symposia Volume: 154, Issue: 1. Date: April, 2000, 73

[7]   Fornel F.,. Springer Series in Optical Sciences, 2001, 73..

[8]   Kosiakov B.I., Tukhvatulin A.Sh. et. at. Journal of technical physics, 1998, 68,10,.70.

[9]   Nadareishvili L., Sh. Gvatua, Japaridze K., Blagidze Yu. Opticheski Journal. Sankt-Peterburg (Optical Journal, Saint-Petersburg), 1997, 64, 12, 62.

[10]   Koike Y., Ishigure T., Satoh M., Nihei E. *Pure Appl. Opt.* 1998, 7, 201.

[11]   Nadareishvili L., Lekishvili N. Polimernie spedi s gradientom opticheskix svoistv (Polymer areas with gradient of optical properties), Tbilisi, «Lega», 2000.

[12]   Nadareishvili L. Georgian Engineering News, 2001, 7, 73.

[13]   Nadareishvili L., Gvatua Sh., et. al. Mezhdynarodnaia konferencia "Prikladnaia Optika" (International Conference: "Applied Optics-2000"). Saint-Petersburg, 2000, p. 10.

[14]   Nadareishvili L., Lekishvili N., Gvatua Sh., Japaridze K. GB-optics–a new direction of gradient optics. In book: Chemistry and Biochemistry on the Leading Edge. Nova Science Publishers, Ed . Zaikov G.E.. New York, 2002, p. 31.

[15]   Nadareishvili L., Gvatua Sh., Blagidze Yu., et. al. Journal of the Balkan Tribological Association, 2003, 9, 2, 207.

[16]   Nadareishvili L., Gvatua Sh., Blagidze Yu., et. al. J. Appl. Pol. Sci. 2004, 91, 489.

[17]   Nadareishvili L., Gvatua Sh., Blagidze Yu., et. al. Opticheski jurnal (Optical Journal), Sanit-Petersburg, 2005, 72, 10, 12.

[18]   Lekishvili N., Nadareishvili L. and Zaikov G. In book: Chemical and Physical Properties of Polymers. Eds: Zaikov G. E. and Kozlowski R.: Nova Science Publishers, Inc. New York, 2005, p.1.

[19]   Lekishvili N., Nadareishvili L., Zaikov G. In book: Modern Advanced in organic and inorganic chemistry. Nova Science Publishers, 2006, New York, p. 31. .

[20]   Weigert F. Verhandlungen Deutschen Physik 1919, 21, 479,

[21]   Zaidel A.N., Ostrovsakaia G.V., Ostrovski Y.N., Texnika i praktika spektroskopii. (The Technique and Practice in the Spectroscopy). Moscow, "NAUKA", 1976, p. 392.

In: Polymer Yearbook – 2011.
Editors: G. Zaikov, C. Sirghie et al. pp. 195-205

ISBN 978-1-61209-645-2
© 2011 Nova Science Publishers, Inc.

Chapter 19

CATALYTIC OXIDATION OF ETHYLENE GLYCOL BY DIOXYGEN IN ALKALINE MEDIUM. THE NEW EXAMPLE OF ONE-STAGE OXIDATIVE CLEAVAGE OF C-C BOND

A. M. Sakharov, P. A. Sakharov and G. E. Zaikov

Institute of Biochemical Physics of N.M. Emanuel
of the Russian Academy of Sciences

ABSTRACT

Reaction of low temperature oxidations of ethylene glycol (EG) by molecular oxygen in the presence of salts of bivalent copper and alkali both in water and in waterless solutions was investigated. It was found that at low (close to room) temperatures and the process carrying out in waterless solutions the basic product of EG oxidation is formic acid. Rise the temperature from $20 – 40°C$ to $80 – 90°C$ or carrying out the reaction in the water-containing solutions leads to sharp change of a direction of reaction. EG in these conditions is oxidized with primary formation of glycolic acids salts. Change of a direction of reaction is connected, apparently, with decrease stability of chelate complexes of Cu^{2+}-ions with dianionic form of EG. The mechanism of glycolic acids formation includes, possibly, a stage of two-electronic reduction of O_2 in reaction of dioxygen with monoanionic forms of EG, coordinated on Cu^{2+}-centers.

Keywords: oxidation by oxygen, ethylene glycol, Cu^{2+}-ions catalysis, formic and glycolic acids formations.

INTRODUCTION

The chemical industry is high on the list among sources of dangerous for environment. In this connection search of the new chemical processes, different low level consumption of energy and minimum formation of by-products is especially necessary. It concerns in the first place to processes of organic substances oxidation by oxygen which allows to produce a wide spectrum of products for various industries. The leader of the works devoted the creation of new high selective processes of liquid-face of oxidation by oxygen was academician N.M. Emanuel. He was sure that the main directions for increasing in selectivity of reactions of liquid-phase oxidation is the use of metal complexes as the catalysts. In the present work kinetic regularities of EG autooxidation in the presence of Cu^{2+}-ions and the bases are investigated. As it will be shown more low, changes of conditions of carrying out of process allows to pass from highly effective oxidation EG to formic acid to reaction with primary formation of glycolic acid. It was found that at optimal conditions the rates of low temperature oxidations of EG close to rates of enzymatic reactions catalyzed by dioxygenases.

EXPERIMENTAL

Oxidation of EG by oxygen in various solvents at 30 - 90°C was spent in the glass reactor equipped with a mechanical stirrer with continues supply of oxygen. At oxygen elevated pressures (to 1 MPa) reaction was carried out in the steel reactor (volume 0.1l). In both cases stirring was carried out by means of mechanical mixers (~ 1000 rotation/min).

Reactionary mixtures was preparing by consecutive introduction of salts of bivalent copper ($CuCl_2.2H_2O$, $CuSO_4.5H_2O$ or $Cu(Ac)_2.2H_2O$) and alkalines (NaOH or KOH) in water or waterless solutions of EG. Alkali addition to solutions of salts of the copper, containing EG, leads to formation of brightly dark blue complexes Cu^{2+}, stable at pH $8 \div 14$. After addition of alkali the reactor was carefully blow of oxygen and pressurized. Oxygen absorption starts after the beginning of stirring of the reaction mix.

Rate of reaction was measured on rates of oxygen absorption , an alkali and EG expenditure and accumulation of products of oxidation of EG. Acids formed in the course of reaction were analyzed by HPLC method on «Millipore Waters» with use UV detector. A column -ZORBAX SAX (4.66 x 250 mm) eluent $\div$ 0.025 M solution KH_2PO_4. Concentration of EG was determined by GLC method [1].

RESULTS

In neutral solutions EG do not react with dioxygen with measurable rates at low 100°C . Alkaline solutions EG in the absence of copper salts do not oxidized by oxygen at low temperatures too. Thus, as well as at oxidation primary and secondary alcohols and polyols [2-5], oxidation of EG at low temperatures by oxygen is realized only in the presence of two-componential catalytic system $\{Cu^{2+} + base\}$.

Practically at once after alkali introductions in a solution containing EG and salt of copper, intensive absorption of oxygen starts. As well as at oxidation of others hydroxy-containing compounds in the presence of copper ions and basis [6], activation of EG in relation to oxygen obviously occurs only after substrate transfer in active anionic form.

The kinetic curves of oxygen absorption at the oxidation of 20 % solutions of EG in tert-butyl alcohol at 50°C in the presence of 5.10^{-3} M $CuCl_2.2H_2O$ and various quantities [KOH] are shown on the Figure1.

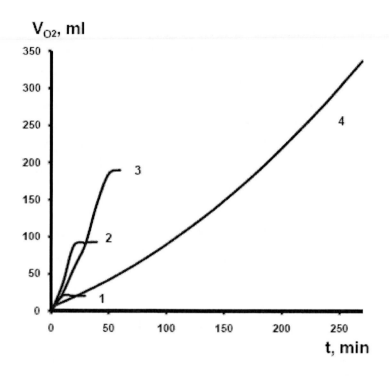

Figure1. Kinetic curve of oxygen absorption in reaction of EG oxidation in the presence of various quantities of the granulated alkali. C_{KOH} (g/l): (1) – 4 g/l; (2) – 14 g/l; (3) – 25 g/l; (4) – 50 g/l.; V = 25 ml, $[CuCl_2.2H_2O]$ = 5.10^{-3}M, P_{O2}=0.1 MPa, $[.EG]_0$ = 200 g/l, solvent- tert-butyl alcohol, 50°C.

It can be seen from Figure1 in the range of KOH concentration 4÷25 g/l (0.07 ÷0.45M) initial rate of absorption of oxygen practically doesn't depend on concentration of alkali (Figure1 curves 1÷3). At higher KOH concentration the rate of oxygen absorption considerably decreases (Figure 1, curve 4) that is connected, apparently, with deactivation of the catalyst owing to formation of inactive hydroxy-complexes of bivalent copper (blue color) in the presence of the high surplus of alkali. Similar kinetic laws of oxidation EG are received at use of NaOH as basis. The maximum rates of oxidation in this case are close to rates of reaction at presence of KOH. However the small period of auto acceleration because of slower dissolution of granules NaOH in comparison with KOH in solutions was observed.

During of EG oxidation in the presence of copper salts and alkali the main products are acids. After alkali neutralization by acids oxygen absorption completely stops (Figure1 curves 1 - 3). Introduction of a new portion of alkali completely restores initial rate of oxygen absorption. The variation of total of the entered alkali allows regulate of depth of EG oxidation. Use of a method of introduction of fractional additives of alkali on a reaction course allows to spend reaction to very high depths of EG transformation (over 95%) at the highest rates of process.

Rate of oxygen absorption at EG oxidation both in water and waterless solutions linearly grows with growth of Cu^{2+}- salts concentration in the range from 0 to $2.10^{-2}M$. Further increase the copper ions concentration do not leads to change of rate of reaction. It is connected, apparently, that in these conditions (high concentration Cu^{2+}) a limiting stage of reaction is the rate of deprotonation of EG.

On Figure2 the dependence of the rate of oxygen consumption on initial EG concentration is presented. Rise of EG concentration in a solution of tert-butyl alcohol from 0 to 50% leads to linearly increase in rate of oxygen absorption. The further increase of EG concentration, leads to considerable decrease in rate of process. It is connected, apparently, with decreasing of donor properties of anion form of EG. Earlier it was be shown [7] that aliphatic alcohols are oxidized with the highest rates only in aprotic solvents. Similar influence of the nature of solvent used on the rate of EG oxidation is observed too.

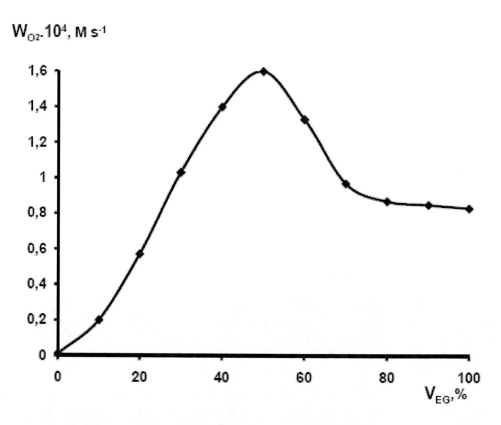

Figure2. Dependence of the rate of EG oxidation from its concentration (vol. %) in a solution of tert-butanol. $[CuCl_2.2H_2O] = 5.10^{-3}M$, $[KOH]_o = 0.5M$, $P_{O2} = 0.1MPa$, solvent tert-butanol, $50^{\circ}C$.

As it is possible to see from Figure3 the highest rates of oxygen combustion in reaction of EG oxidation are reached if the most polar aprotic solvents, such as DMSO (column 1) and DMF (column 2) are used. At the use of H-containing solvents rate of reaction are decreasing and in water containing solvent (column 8) makes not more than 2-3% from the value of rate of EG oxidation in DMSO.

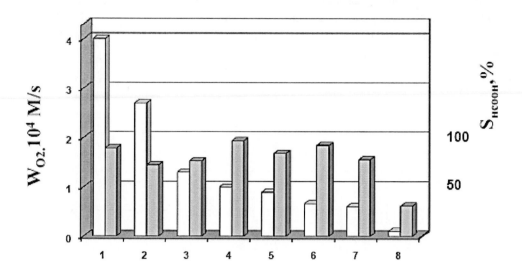

Figure 3. Rates of oxygen absorption (WO2 ,white bar) and selectivity of formic acid formation (S, grey bar) □ at EG oxidation in various solvents: 1 – DMSO, 2 – DMF, 3 – ethanol, 4 – EG, 5 – diglym, 6- tert-butanol, 7 – 2-propanol, 8 – water; [CuCl2.2H2O] = 5.10-3M, [KOH]o= 0.5 M, 50oC, [EG] = 20% vol. (1-3; 5-8), [EG] = 100 % vol. (4). PO2 =0.1 MPa.

The analysis of products of EG oxidation by GLC and HPLC methods has shown that in the course of reactions in all investigated conditions as the basic products of reaction acids are formed. Depending on conditions of carrying out of process EG can be oxidized to formic acid (with selectivity close to 100%) or with primary formation glycolic acids.

At optimum conditions (not water containing solutions and low temperatures) salts of formic acid are the unique products of EG oxidation (reaction I):

$$HOCH_2\text{-}CH_2OH + 1.5O_2 + 2KOH \rightarrow 2HCOOK + 3H_2O \qquad (1)$$

On Figure3 along with the rates of absorption of oxygen data, the data on selectivity of formation of formic acid on EG concentration is presented. The greatest selectivity of formation of formic acid, coming nearer to 100%, is observed in case of oxidation pure EG (column 4) and 25% of solutions in tert-butyl alcohol (column 6). Oxidation of water solutions of EG leads to sharp decrease in rate of reaction. In the presence of water, unlike oxidation EG in waterless environments, the basic product of oxidation EG is not formic but glycolic acid (reaction 2).

$$HOCH_2\text{-}CH_2OH + O_2 + KOH \rightarrow HOCH_2\text{-}COOK + 2H_2O \qquad (2)$$

In Table 1 the date on rate of oxygen absorption and selectivity of formic and glycolic acids formation is presented. Increase in concentration of water above 10% in mix of H_2O – tert-butyl alcohol leads to sharp decrease in rate of absorption of oxygen and to considerable increase of concentration glycolic acids in products of EG oxidation.

Table 1.Rates of O_2 absorption (W_{O2}) and selectivity of formic (S_{HCOOH}, %) and glycolic ($S_{HOCH2COOH}$, %) acids formation in solution H_2O – tert-butanol at various concentrations of H_2O in a mix (vol. %).$[CuCl_2.2H_2O]_o$=5.10^{-3}M, [EG] =20% vol., $[KOH]_0$ =0.5M, T=50^oC, P_{O2} =0.1MPa

Contents of H_2O in the mix H_2O – tert-butanol, vol %	0	5	10	25	50	100
$W_{O_2}.10^5$,M/s	6.8	6.6	5.9	2.5	1.1	0.5
S_{HCOOH}, %	96	93	91	77	45	40
S_{HOCH_2COOH}, %	1	5	7	19	40	52

In Table 2 the rates of oxygen absorption data at various temperatures are presented. In the range of temperatures from 30 to 70^oC classical Arrhenius dependence is observed: energy activation is 43 kJ/mol. However at rise in temperature above 70^oC rate of absorption of oxygen doesn't increase. It is possible to see that at rise in temperature from 70 to 80^oC in reaction of oxidation of 20% vol. solution of EG in tert-butyl alcohol selectivity of formic acid formation decrease from 80 to 55%. It is connected, apparently, with the rise of a rate oxidation of K-salts of EG to CO_2.

Table 2. Rates of O_2 absorption and selectivity of formation formic and glycolic acids at various temperatures. $[CuCl_2.2H_2O]_o$=5.10^{-3}M, [EG] =20% vol., [KOH] =0.5M, P_{O2}=0.1 MPa., solvent – tert-butanol

Temperature, ^{o}C	30	40	50	60	70	80
$W_{O_2}.10^4$, M/s	0.1	0.33	0.68	1.05	1.7	1.71
S_{HCOOH}, %	98	97	96	91	80	55
S_{HOCH_2COOH}%	0	0.5	1	4	5	12

Oxidation of EG in a bulk at two temperatures - 50 and 90^oC was studied too (Table 3).

Table 3.Rates and selectivity of formic and glycolic acids formation at 50oC and 90oC. [CuCl2.2H2O] = 5.10-3M, [KOH]0 = 0.4M, PO2 =0.1MPa.100% vol. EG

T ^{o}C	$W_{O_2}.10^4$ M.s^{-1}	S_{HCOOH}, %	S_{HOCH_2COOH}, %
50	1.0	95.0	3.5
90	0.9	35.0	55.0

As can be seen from this table, rise in temperature on $40^{\circ}C$ leads to decrease in rate of oxygen absorption and to change of a direction of process. If at low temperatures almost unique product of oxidation is formic acid, at $90^{\circ}C$ EG is oxidized mainly to glycolic acids.

Dependence of rate of oxidation EG from partial pressure of oxygen is unusual to reactions of liquid phase oxidations of organic connections. In Table 4 the data of influence on rate and selectivity of EG oxidation to formic acid in tert-butyl alcohol from O_2 pressure are presented.

**Table 4. Rates and selectivity of formic acid formation
at various pressure of oxygen.Solvent tert-butanol, EG 25% vol,
$[CuCl_2.2H_2O] = 5.10^{-3}M$, $[KOH]_0 = 0.5M$, $50^{\circ}C$**

Partial presser of O_2	Rate of KOH consumption $M.s^{-1}$	Selectivity of HCOOH formation, %
0.02 MPa.	$2.1.10^{-5}$	45
0.1 MPa.	$6.3.10^{-5}$	90
1.0 MPa.	$3.2.10^{-4}$	≈ 100

Selectivity of formic acid formation was determent as the relation of concentrations of HCOOH formed to concentration of hydroxide potassium spent. It is well known that the rate of liquid phase oxidations of organic substrates in the presence of initiators or traditional catalytic systems in neutral environments practically do not depend from partial pressure of oxygen at PO2 from above 0.02 MPa [8]. As it is possible to see from Table 4 considerable growth of rate of oxidation EG is observed at least up to values PO2 – 1MPa. At the low pressure of oxygen (0.02 MPa.) not only low rate of oxidation of EG is observed, but low selectivity of formic acid formation too. At elevated pressures of O2 and at low temperatures the unique product of reaction is HCOOH.

DISCUSSION OF RESULTS

The kinetic regularities of EG oxidation by O_2 resulted in an experimental part demonstrate that in the presence of ions of bivalent copper and alkali reaction proceeds via the uncommon mechanism. Extraordinary strong dependence of the rate and the direction of EG oxidation on the nature of solvent used was found. Dependence of rate and selectivity of oxidation on temperature is uncommon too. Besides that dependence on rate of reaction from partial pressure of oxygen is atypical for reactions of liquid phase oxidations of organic substrates by oxygen.

On the basis of the received experimental data it is possible to draw a conclusion that in most cases (waterless environments, moderate temperatures) oxidation proceeds through the stage of chelate complexes formation between Cu^{2+}-ions and dianionic forms of EG. Dark blue complexes of bivalent copper with dianionic form of EG are formed practically immediately after alkali additions to solution containing of EG and Cu^{2+}salts. Such complexes possess very high stability. They have been allocated from solutions and characterized by a method X-ray analysis [9].

It is known that bivalent copper ions can catalyze oxidation by molecular oxygen of some easily oxidized compounds such as phenols, pyrocatechol, etc. which are in anionic form. In such reactions, as it has been shown [10], bivalent copper ions act as oxidants and the role of oxygen consists in reoxidation of unstable Cu^+ cations, formed on the first stage of reaction, to Cu^{2+} ions. Differ of this, Cu^{2+} complexes with dianionic form of EG in the absence of oxygen are stable, that can be confirmed by spectral analysis of such complexes. Thus, the mechanism of the reaction which suppose the stage of EG ions oxidation by bivalent copper ions (reaction 3) must be excluded from consideration:

$$Cu^{2+} ... ^-OCH_2CH_2O^- \longrightarrow //\rightarrow Cu^+ ....OCH_2CH_2O^- \tag{3}$$

It was possible to assume that as active intermediates in reaction of EG oxidation are the reduced forms of oxygen, such as superoxide-anion radical or hydrogen peroxide. It is well known, that superoxide-anion radicals are absolutely unstable at presence even water traces. But, as it is possible to see from Table 1, introduction in a solution from 5 to 10% vol. water practically doesn't influence on rate of oxygen consumption. .

On Figure4 the kinetic curve of oxygen absorption in the reaction of EG oxidation in tert-butyl alcohol at introduction in a reactionary mix of additives of 30% of solution H_2O_2 is presented. One can see that after introductions of H_2O_2 in solution an intensive allocation of oxygen is observed . During 3 ÷5 minutes almost all entered hydrogen peroxide decompose on oxygen and H_2O. The products of EG oxidation are almost similar to that as in without H_2O_2 addition.

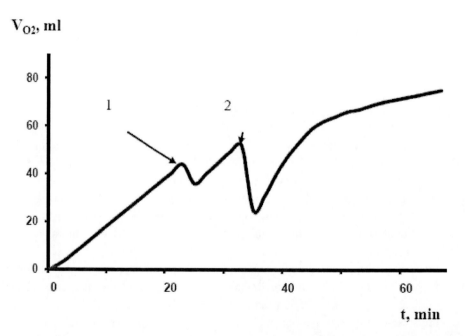

Figure 4. The kinetic curve of oxygen consumption in reaction of EG oxidation (20% vol.) in tert-butyl alcohol. Arrows specify the moment of introduction in the reactionary mix of 30% of solution H2O2 in quantity: (1) – 1.9 mmol, (2) – 3.8 mmol H2O2; V= 25 ml, [CuCl2.2H2O] = 5.10-3 M, [KOH]o = 0.3M, 50oC, PO2 =0.1 MPa.

As it was specified above, numerous reactions of primary and secondary alcohols, glycols, ketones in the presence of complexes of copper and the bases proceeds, apparently, via thermodynamic favorable multi electronic mechanism of carrying over electrons from anionic forms of substrates, coordinated on Cu^{2+} the centers, on O_2 molecules [1-7]. Efficiency of direct interaction of coordinated anions with O_2 is in many respects defined their donor by ability. The results on influence of rates of EG oxidation by nature solvent confirm this. The highest rates of oxidation are observed when the process carrying out in DMSO or DMF (Figure3), i.e. when the solvents in which donating ability anions the highest are used [11]. At process carrying out in the water, owning the most strong proton donating ability, exceptional drop in a rate of reaction is observed.

High stability of complexes of copper with anionic form of EG to redox transformations in the absence of oxygen and unusually strong dependence of rate of reaction to partial pressure of oxygen, confirm that direct attack O_2 to a complex of bivalent copper with anionic form of EG is limiting stage of reaction.

The received kinetic dates of oxidation of EG allow to assume that formation of formic acid results from interaction O_2 with chelate complexes $\{Cu^{2+} ... A^{2-}\}$ (where A^{2-} dianionic form of EG). Stability of such complexes in alkaline solutions, as it was already marked, very high. Increase of concentration of anions Cu^{2+} leads to increase of concentration of such complexes and linear increase of rate of oxygen consumption of and the rate of formic acid formation.

The structure of copper complexes with dianionic form of EG in very strong degree depends of temperature and nature of solvent used. It is known, that stability of chelate complexes of transition metals decreases at rise in temperature. Replacement aprotic solvents on water also should lead to decomposition of copper complexes with EG-dianions due to water, as known, is the most powerful hydrolyzing agent.

On the basis of the represented experimental data it is possible to assume that the first stage of the reaction leading to formation of formic acid is the interaction of $\{Cu^{2+} ... A^{2-}\}$ complex with O_2, where A^{2-} - dianionic form of EG. Abnormal strong dependence of rate of reaction on partial pressure of oxygen conforms this supposition. It was noticed that complexes Cu^{2+} with anionic form of EG are very stable in anaerobic conditions (reduction of Cu^{2+} ions to Cu^+ do not proceed with measurable rates).

It is can be assume that formic acid is the product of subsequent oxidation of primary product - glycolic acid. But introduction glycolic acid in a EG solution (in the presence of 5.10^{-3} $CuCl_2.2H_2O$, 0.2 M of glycolic acids and 0.5M KOH in tert-butyl alcohol, $50^{\circ}C$) do not change the rate of process and rate of formic acid accumulation. It is not conformed with the assumption of participation glycolic acids as intermediate in HCOOH formation.

Oxidizing rupture of C-C bonds in EG molecule can leads to methanol or formaldehyde formation, also possible intermediates in formic acid formation. However, methanol even as traces was not detected in products of EG oxidation.

Earlier it was shown, that formaldehyde which could be formed as intermediate compound, in alkaline environments in the presence of copper ions is oxidized with high rates, but with very low selectivity of formic acid formation [12]. Reaction of EG oxidation to formic acid proceed, apparently, via the concert mechanism and includes multi-electronic reduction of oxygen without formation stable intermediates. It is known, that such processes with participation O_2 as an oxidizer is most favorable from the thermodynamic point of view [13]. The participation Cu^{2+} ions in the process of electrons transfer from coordinated anion form of substrate to O_2 molecule open possibility of reaction proceeding by concert multi-electron mechanism without formation of free radicals. Proceeding of process by thermodynamically advantage multi-electron mechanism of oxygen reduction allows reach of extremely high rates of oxidation at room temperatures at selectivity of formation of final products exceeding 90%.

CONCLUSIONS

The resulted experimental data show that at use of oxygen as the cheapest and non-polluting oxidizer and simple catalytic system [Cu^{2+} ... substrate ... OH⁻] is possible to oxidize with extraordinary rates even such inert in relation to O_2 organic compounds as EG. It was shown, that varying conditions of carrying out of process it is possible to obtain as a main product glycolic or formic acid with high selectivity. The rates of processes are near to that of enzymatic reactions at the presence of dioxygenases.

REFERENCES

[1] Sakharov A.M., Mazaletskaya L.I., Skibida I.P. // Catalytic Oxidative Deformylation of Polyethylene Glycols with the Participation of Molecular Oxygen // *Kinet.Catal.*, 2001,vol 42, № 5, pp. 662-668.

[2] Skibida I.P., Sakharov A.M. // Chemical model of oxidases. Cu^I-catalyzed oxidation of secondary alcohols by dioxygen// *Russian Chemical Bulletin,* 1995, vol. 44, No 10, pp. 1872- 1878.

[3] Sakharov A.M., Skibida I.P. // Novel chemical models of enzymatic oxidation. II Oxidation of alcohols to acids in the presents of Cu(II) complexes // Kinet.Katal., 1988, vol.48, no 1, p. 157.

[4] Sakharov A.M., Silakhtaryan N.T., Skibida I.P. //Catalytic Oxidation of Polyols by Molecular Oxygen in Alkaline Media // *Kinet.Katal.,* 1996, vol.37, No 3, pp. 368-376.

[5] Sakharov A.M., Skibida I.P. // Mechanism of catalytic oxidation of polysaccharides by molecular oxygen in alkaline gels // *Chem. Ph.,* 2001, vol.20, p. 101. (in Russian).

[6] Skibida I.P., Sakharov A.M. //Molecular oxygen as environmental acceptable, selective and the most strong oxidany in liquid-phase oxidation //Catalysis Today. 1996. vol.27, p.187 - 193.

[7] Sakharov A.M., Skibida I.P. //Oxidative Cleavage of yhe C-C bond in the Catalytic Oxidation of Alcohols with Dioxygen//*Dokl.Phys.Chem.,* 2000. vol. 372. No 6, pp. 785-788.

[8] Emanuel N.M., Denisov E.t., Maizus Z.K. //Liquid-Phase Oxidation of Hydrocarbons.// (Plenum Press. New York), 1967.

[9] Habermann N., Jung G., Klaassen M., Klüfers P. // Polyol-Metall-Komplexe, II Kupfer(II)-Komplexe mit mehrfach deprotoniertem Ethylenglycol, Anhydro-erythrit oder Methyl-α-D-mannopyranosid als Liganden // *Chem.Ber.*1992. V.125. P.809-814.

[10] Demmin T.R., Swerdloff M.D., Rogic M.M. // Copper(II)-induced oxidations of aromatic substrates // J.Amer.Chem.Sos. 1981. V.103. P.5795-5084.*J.Am.Chem.Soc.* 1981. vol.103. p.5795-5884.

[11] Russell G.A., Bemis A.G., Geels E.J., Moye A.J., Jansen E.G., Mak S., Storm E.T. // Oxidation of Hydrocarbons in Basic Solution// *Adv.Chem.Ser.* 1965. V.51. P.112.

[12] Skibida I.P., Sakharov A.M. // Perspectives in environment of selectivity in liquid phase oxidation by dioxygen. New models of enzymatic oxidation. // New Developments in Selective Oxidation. Centi G., Trifiro F. (Editors). 1990. Elsevier Science Publishers B.V., Amsterdam. P.221- 228.

[13] Jones R.D., Summervile D.A., Basolo F. // Synthetic oxygen carriers related to biological systems // *Chem.Rev.* 1975. V.79. № 2. P.139-179.

In: Polymer Yearbook – 2011.   ISBN 978-1-61209-645-2
Editors: G. Zaikov, C. Sirghie et al. pp. 207-219   © 2011 Nova Science Publishers, Inc.

Chapter 20

THE ANALYSIS OF CHANGES OF RELAXATION PARAMETERS DURING THEIR MEASUREMENT

*N. N. Komova and G. E. Zaikov**

Moscow State Academy of Fine Chemical Technology,[1]
Moscow 119571, Russia
*N.M. Emanuel Institute of Biochemical Physics Russian Academy of Sciences[2]
Moscow 119334, Russia

ABSTRACT

It is shown, that at research such relaxation characteristics as the tangent of mechanical losses, dissipation of mechanical energy as a result of the internal friction, measured at periodic action of monoaxial compressive stress on the sample of polymer (LDPE) in the high elasticity state, occurs development relaxation processes in the sample. In this connection it is necessary to take into account the temperature-time conditions of the experiment or introduce appropriate amendments to the results obtained.

Keywords: relaxation processes, dissipation mechanical energy, a tangent of mechanical losses, a principle of temperature-time superposition.

INTRODUCTION

Any measurement of physical system is made by means of some device (in the more general case - the measuring environment). Thus there is an interaction of the device to measured system therefore the system condition to some extent changes depending on intensity of influence from the device. At measurements of classical system quite pertinently to suppose, that the measurement doesn't change a condition of measured system at all. If

---

[1] 124 Vernadsky avenue, Moscow 119571, Russia, komova_@mail.ru.
[2] 4, Kosygin st., Moscow 119334, Russia, chembio@sky.chph.ras.ru.

describe a condition of measured system and measurement procedure make so in details that features of influence of measuring system that a situation cardinally are shown changes. It appears that owing to the quantum nature of physical systems during the measurement the conditions of measured system are changing. These changes are so more than more information received by measurement. It is necessary to pay for the information. So in the theory of measurements the information increase corresponds to reduction of entropy [1]:

$$S = -\sum_{i}^{n} p_i \ln p_i,$$

where p_i - aprioristic probabilities of various conditions of system, n – quantity of conditions.

Thus, increasing accuracy of measurement, we necessarily increase also return influence of measuring procedure by a condition of measured system.

John Neumann for quantum system has proved a background and mathematically has strictly formulated a postulate of a reduction. According to this postulate at measurement of some observable size the system condition changes in such a manner that in a new condition the measured observable has already another certain value, and it has turned out as a result of measurement. Occurrence of this condition is called as a condition reduction of a system.

In the theory of measurements it is considered two types of measuring systems: passive and active [2]. In the passive measuring system there is a comparison of the defined size with the standard without any active influence on the system which parameters define. The feature of active measuring system is influence on characterized system, and the response of system to this influence gives the information for calculation of demanded parameters.

As the active measuring system assumes a certain influence on characterized object in the course of this influence the object can undergo changes. Therefore for reception of the most exact value of the defined parameter in the theory of measurements perform the operation of coordination between measuring system and the measured object, consisting in reduction, and at the best data dissipation, influences of entrance influence on measured object.

At the measurements concerning difficult systems or objects, the measured size often depends on set of various circumstances. Usually the nature and quantitative characteristics of these dependences are unknown. The circumstances influencing on result of measurement don't remain constants during carrying out of measurement, there fore it's impossible to correct this or that error of measurement. It means that measurement isn't selective, and the result of measurement comprises also other factors.

In the greatest measure these principles are important for sizes which assume measurements in which basis the difficult physical and mathematical models demanding a certain sort of updating according to conditions of measurements lie. In the mechanic of polymers such sizes are the parameters characterizing relaxation properties of materials. These sizes and dependences corresponding to them give the chance to judge structure of polymers, to find temperatures of structural transitions and service conditions of corresponding materials [3,4].

THEORETICAL PART

One of widely used methods in research elastic and relaxation properties of polymers in the block at periodic sinusoidal loadings is the method of Aleksandrova-Lazurkina [6]. Unlike resonant this method is applied for high elasticity deformations of polymers in the field of the frequencies lying considerably below own frequency of the sample – far from resonant area. In this case, the phase relations, phase lag of deformation from stress-relaxation time is determined only by or through the appropriate range of relaxation time and elastic material. In this method phase parities don't depend on the form, the size and density of the sample that allows to find relaxation time of a material from measurements .

The method is based idea of rubbery (*high elasticity*)deformation as a reflection of the deformation of flexible macromolecules, and the appearance of the elastic forces of deformation and shape recovery after unloading - the result of thermal motion of parts of macromolecules. However, all the patterns that underlie the method, refer to the equilibrium states of the body under load. The study of temporal patterns of rubbery deformation in a regime of constant stress or strain, as well as in periodic loads confirmed the significant role of the kinetics of rubbery deformation relaxation phenomena in the behavior of polymeric materials under mechanical stress, and in the process of vitrification of polymers [5,7].

Depending on the interim regime changes impact the behavior of the material. At a constant temperature with increasing speed or increasing the frequency of impacts observed so-called effect of "hardening" of the material [8].

Total deformation of the polymer is composed of an elastic, rubbery (high elastic) flow and deformation. When considering the polymer in the rubberlike (high elasticity) state accepts that the macroscopic viscosity of the material is great and flow absent.

To obtain the dependence of rubbery component of the strain on the applied stress using the simplest model for these conditions[9]. In this case, such a model is a three-element model representing a model of Kelvin (parallel connected spring and damper), connected in series with a spring. The equation describing the relationship between stress and strain of this model is as follows:

$$\frac{d\sigma}{dt} + \frac{E_0 + E_1}{\eta}\sigma = E_0\frac{d\varepsilon}{dt} + \frac{E_0 E_1}{\eta}\varepsilon, \tag{1}$$

where σ -stress acting on the system being studied; ε-deformation occurring in the system under the applied stress, E_0- the module of elasticity; E_1- high elasticity module; η – micro viscosity.

Then the deformation of the polymer is made up of elastic strain $\varepsilon_0 = \sigma/E_0$ and high elasticity ε_1 parts. Rewriting equation (1) relative rates of change of strain and isolating highly elastic component of deformation, we can received :

$$\frac{d\varepsilon_1}{dt} + \frac{E_1}{\eta}\varepsilon_1 = \frac{\sigma}{\eta} \tag{2}$$

If the stress varies with time harmonically with frequency ω:

$$\sigma = \sigma_0 \cos \omega t \, , \tag{3}$$

full deformation is described by the equation:

$$\varepsilon = C e^{-\frac{t}{\tau}} + \sigma_0 \left\{ \left(\frac{1}{E_0} + \frac{1}{E_1} \frac{1}{1+\omega^2\tau^2} \right) \cos \omega t + \frac{1}{E_0} \frac{\omega\tau}{1+\omega^2\tau^2} \sin \omega t \right\} , \tag{4}$$

where the parameter $\tau = \eta / E_1$ is called as *relaxation time*. In some works [10,11] this parameter is named delay time, and relaxation time is named the parameter, which proportional to it:

$$\tau_1 = \tau \frac{E_1}{E_1 + E_0} \tag{5}$$

The first exponential member of the equation (4) contains constant C, witch depending on initial conditions, and defines an unsteady part of deformation fading in due course. Therefore, if from the beginning of carrying out of measurement has passed enough time t>> τ (the transients which have arisen at the moment of a start of motion, have already faded and takes place the established conditions) it is possible to neglect this member and consider only that part of expression (4), which is concluded in braces. This part describes the stationary oscillations, which are studied on experience. They consist of oscillations in a phase with the pressure, described by expression in parentheses and consisting of elastic and *high elastic* components, and the oscillations, which are lagging behind pressure on a phase on π/2. As these two harmonious oscillations are directed along one axis (a vector of their speeds are collinear) the amplitude of deformation is expressed by the equation:

$$\varepsilon_0 = \sigma_0 \sqrt{ \left(\frac{1}{E_0} + \frac{1}{E_1} \frac{1}{1+\omega^2\tau^2} \right)^2 + \frac{1}{E_1^2} \frac{\omega^2\tau^2}{\left(1+\omega^2\tau^2\right)^2} } \tag{6}$$

Using condition $E_0 >> E_1$: as the high elasticity module for polymeric materials on some orders less than the module of elasticity, it is possible to receive dependence of deformation on pressure and frequency of loading (ω):

$$\varepsilon_0 = \frac{\sigma_0}{E_1} \frac{1}{\sqrt{1+\omega^2\tau^2}} \tag{7}$$

The received expression can be transformed as:

$$\frac{\varepsilon_0}{\sigma_0} = \frac{1}{E_1 \sqrt{1+\omega^2\tau^2}} \tag{8}$$

Parameter ε_0/σ_0 is dynamic compliance (I) and equal to the inverse dynamic module. The compliance makes sense strain in a single strain.

Using complex representation of harmoniously changing deformation: $\varepsilon(t) = \varepsilon_0 e^{i\omega t}$, strain rate will have an expression $d\varepsilon(t)/dt = \omega \varepsilon_0 e^{i(\omega t + \pi/2)}$. Substituting this expression in the differential equation (1) and reducing on $\varepsilon_0 e^{i\omega t}$, we received:

$$(i\eta\omega + E)E^*(i\omega) = iE\eta\omega , \tag{9}$$

where E^* is a complex dynamic modulus, which can be represented as:

$$E^*(i\omega) = \frac{\eta^2\omega^2 E}{E^2 + \eta^2\omega^2} + i\frac{\eta\omega E^2}{E^2 + \eta^2\omega^2} \tag{10}$$

The first term is a real, and the second - the imaginary part of the complex dynamic modulus ($E^* = E' + iE''$), which is proportional to E and depends on the frequency . $E''(\omega)$ determines the losses at harmonic deformation and is the module of losses.

Similarly complex dynamic module $E^*(i\omega)$ can be represented by a complex dynamic compliance I^* as the sum of the imaginary and real I'' I' parts. Considering that $I^*(i\omega) E^*(i\omega) = 1$, we can provide the relevant expressions in the form: $I^*(i\omega) = I'(\omega) + i I''(\omega)$, where

$$I' = \frac{1}{E_0} + \frac{1}{E_1}\frac{1}{1+\omega^2\tau^2} = I_0 + \frac{I_1}{1+\omega^2\tau^2} \tag{10}$$

$$I'' = \frac{1}{E_1}\frac{\omega\tau}{1+\omega^2\tau^2} = I_1\frac{\omega\tau}{1+\omega^2\tau^2} , \tag{11}$$

where $I_0 = 1/E_0$, and $I_1 = 1/E_1$. Absolute measured deformation looks like: $I = \sqrt{I'^2 + I''^2}$
From condition $I_1 \gg I_0$ (as $E_0 \gg E_1$):

$$I \approx \frac{I_1}{\sqrt{1+\omega^2\tau^2}} \tag{12}$$

The phase angle δ between the I and I ``, i.e between strain and stress is defined as:

$$tg\delta = \frac{I''}{I'} = \frac{I_1 \omega \tau}{I_1 + I_0(1 + \omega^2 \tau^2)} \qquad (13)$$

In essence the angle δ describes the mechanical loss, i. e share of mechanical energy, which came into heat, or the proportion of dissipated energy per cycle of deformation per unit volume. A measure of this transformation may be an area corresponding to the hysteresis loop formed by the dependence of deformation on the voltage in the cycle of periodic actions (between the curve of loading and unloading).

At low frequencies, when you can measure the hysteresis loop and hysteresis loss coefficient is used mechanical loss [3]: $\chi = \Delta W/W$, where W-total work force for a series of mechanical deformation, and ΔW - dissipated energy per cycle of deformation, which is proportional to the square hysteresis loop.

Between χ and tgδ there is a dependence at all frequencies in terms of linear viscoelasticity Thus, for asymmetric vibrations from 0 to $2\varepsilon_0$ according to work [12] such dependence is found:

$$\chi = \frac{2\pi tg\delta}{4\sqrt{1 + tg^2\delta} + \pi tg\delta} \qquad (14)$$

The decision of this equation concerning parameter tg δ gives dependence:

$$tg\delta = \frac{4\chi}{\sqrt{4\pi^2(1-\chi) - 6\chi^2}} \qquad (15)$$

This expression can be represented as: tgδ = ψ. In expression (13) from condition $E_0 \gg E$ and $I_1 \gg I_0$ at low frequencies in a first approximation, we obtain:

tgδ= $\omega\tau$ (16)

Equating last two expressions, we receive: ψ = $\omega\tau$, whence τ = ψ / ω.

According to the second postulate of the Boltzmann adopted in his theory of the elastic aftereffect, and the underlying Boltzmann-Volterra model that describes the relaxation phenomena, using a function of heredity [13]: action occurred in the past few strains on the stresses caused by deformation of the body at any given time, do not depend on each other and therefore algebraically added. This position has received also the name of a principle of the Boltzmann`s superposition . It should be noted that the polymer body superposition principle holds in the upper-bounded the range of deformation, stress and rate of change.

Given this principle, considering the dissipative processes occurring during application of periodic voltage to the material in the rubberlike state for a long time, we can conclude that there is an accumulation of mechanical energy dissipation in each cycle. Then, if the energy is transformed into heat during one cycle is determined by parameter χ_1, then in a low heat with the environment during N cycles of the energy dissipated during the time t will be: $\chi_{com} = \chi_1 t$ v, where $t\,v = N$.

Stored energy in the sample is converted into heat, which should lead to an increase in temperature. The principle of temperature-time superposition [14], which establishes the equivalence of the effect of temperature and duration of exposure on the relaxation properties of polymers, we can assume that the increase in the impact load on the material is proportional to the action of temperature. Empirical dependence of the temperature ΔT of the exposure time t and the intensity (frequency) exposure to v in the first approximation be written as: $\Delta T = bt\,v$, where the b-parameter, taking into account the characteristics of energy conversion, depending on the structure of the material .

Relaxation time of the supplied periodic voltage decreases with increasing temperature and obeys the Arrhenius equation :

$$\tau = \tau_0 e^{U/RT} \tag{17}$$

For elastic-plastic bodies similar dependence follows from Aleksandrova-Gurevich's equation [15] and has the form

$$\tau = \tau_0 \exp[(U_0 - a\sigma)/RT], \tag{18}$$

where U_0 - activation energy of relaxation process, the constant of the material.

This equation takes into account the dependence of the relaxation of the load. If we assume that $U_0 - a\sigma \approx U$ and to determine the relative relaxation time as τ_t/τ_1 (the ratio of the current value of the relaxation time of the initial value of you during the load application), then, on the basis of equation (18), we can represent this value as an expression:

$$\frac{\tau_t}{\tau_1} = \exp\left(\frac{U}{R(T_1 + \Delta T)} - \frac{U}{RT_1} \right) \tag{19}$$

where the temperature T_1 corresponds to the beginning of load application in the relaxation time τ_1, and the increment of the ΔT is the temperature change in the impact load..

After application of rather simple algebraic transformations the formula (19) will become:

$$\frac{1}{\ln\dfrac{\tau_1}{\tau_t}} = \frac{RT_1}{U} + \frac{RT_1^2}{U}\Delta T . \tag{20}$$

If instead the increment of temperature ΔT would use the proposed higher proportion of the value of the exposure time t and frequency of the applied load v, then the expression (20) becomes:

$$\left(\ln\frac{\tau_1}{\tau_t} \right)^{-1} = \frac{RT_1}{U} + \frac{RT_1^2}{U}(b\,vt)^{-1} \tag{21}$$

Using this expression and taking into account the approach adopted, it is possible to experimental data on changes in the mechanical loss factor (tangent of mechanical losses) over time, the impact loads to find the estimates of the activation energy of relaxation process, determine the extent to which the process is stationary (steady), the degree of linearity of the relaxation processes and the range of conditions and the regime correct determination of relaxation parameters for a periodic load.

EXPERIMENTAL PART

In the present work as object of research polyethylene of low density (*LDPE)*). Samples in the form of the cylinder: diameter (d) from 8 mm at a parity h/d = 1,5 made pressing at temperature 180°C, pressure of 150 kgs/sm². To obtain a homogeneous sample produced an extract of the pressure and temperature 180°C at least 10 minutes with pre-pressing for the release of air located between the grains of the original polymer .

Samples subjected to periodic monoaxial deformation of compression on the installation described in [16] at room temperature (293 K). As a result of the periodic action of the voltage on the sample received the stress-strain during loading and unloading of the form a hysteresis loop. The study used three discrete frequencies of loading: 0.017, 0.17 and 1.7 Hz. Under each of these frequencies is tested at least three samples for 30 min., Taking readings every 5 min. Results cheated, define the parameters of the mechanical losses as the ratio of the hysteresis loop to the area between the curve of loading and the axis of strain ($\chi = \Delta W/W = S_{loop}/S_{hole}$.). The results of measurements of at least three samples were averaged and subjected to further processing in accordance with those presented in the theoretical part of the calculations.

Figure 1 shows kinetic curves of variation of the mechanical loss by prolonged exposure of three frequencies: 0.017, 0.17 and 1.7 Hz. It is seen that with increasing time of deformation coefficient of mechanical losses vary, but relationships have different characteristics

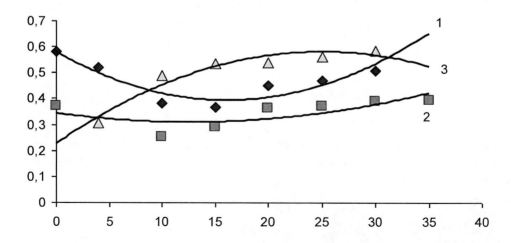

Figure1. Change of the mechanical losses coefficient eventually at influence of periodic loadings with frequency: 1- 0.017 Hz, 2 – 0.17 Hz, 3 – 1.7Hz.

So for low frequencies 0,017Гц (curve 1) and 0,17 Hz (curve 2)the initial value of this parameter is higher than the next. In all probability this is due to the fact that during the reduction of χ is the system output at steady state, ie where the constant C in equation (4) becomes equal to 0. For higher frequency - 1.7 Hz (curve 3) the establishment of this regime is much faster. For higher frequency - 1.7 Hz (curve 3) the establishment of this regime is much faster.

$1/(\ln(\tau_1/\tau_t))$

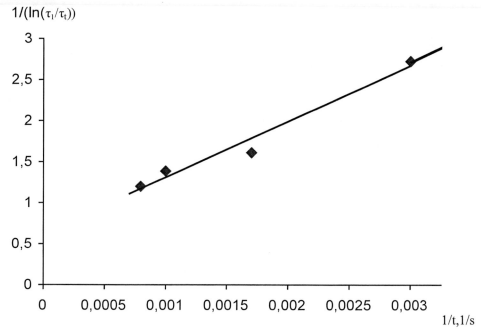

$1/t, 1/s$

Figure 2. Dependence of relative time of a relaxation (τ_t/τ_0) on duration of influence with frequency of 0,017 Hz

Figure 2 shows the inverse of the logarithm of the relative relaxation time in degree -1 (which corresponds to the left side of the equation (21)) the reciprocal of the time of impact load on the sample with a frequency of 0.017 Hz. Dependence is well approximated by a straight line, i.e. the found coordinates dependence of relative time of a relaxation and time of influence of loading is directly proportional. Meaningfully the value found at the intersection of this dependence with the vertical axis and bearing T_1 equal to ambient temperature (293 K) can determine the activation energy. In these conditions (Figure 2) it is equal to 4.9 kJ / mol. The slope in Figure 2 provides an estimate of the value of the parameter "b" in formula (21) . The calculation shows that the frequency of 0.017 Hz, b = 11,88. Since the dependence is linear in a rather wide time interval, this gives reason to conclude that the activation energy of relaxation process with periodic loading of the solid LDPE under these conditions virtually unchanged.

Using the representation of the relative relaxation time of the duration of the periodic effects in the corresponding coordinates for the frequency of 0.17 Hz (Figure 3) makes it possible to calculate the activation energy and the parameter b for the relaxation process of solid LDPE under these conditions. With accuracy to the experimental errors (for a test frequency of 0.17 Hz), the activation energy is 4.9 kJ / mol and b = 0,414.

N. N. Komova and G. E. Zaikov

In Figure4 a similar dependence is shown for the frequency of 1.7 Hz. The calculated value of activation energy is the 2,4 kJ/mol. The value b =0,04.

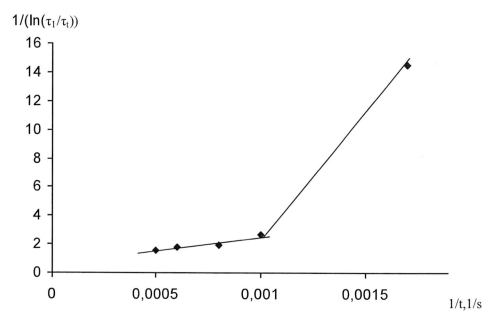

Figure 3. Dependence of relative time of a relaxation (τ t/τ0) on duration of influence with frequency of 0,17 Hz

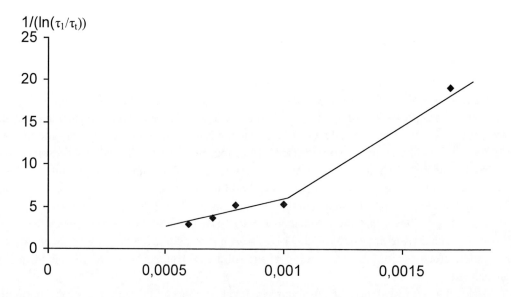

F igure 4. Dependence of relative time of a relaxation (τ $_t$/τ_0) on duration of influence with frequency of 1,7 Hz.

Analyzing the obtained values, we should note a decrease of b with increasing frequency (Figure 5), indicating that the difference in the relaxation processes occurring at different frequencies. Another interesting fact is that for frequencies 1.7 and 0.17 Hz sampling rate on the parameter b is the same and equal to 0.07, while the frequency of 0.017 Hz (three orders of magnitude smaller than the largest) is the product of three times and equals 0,202. It should be noted the difference in the nature of the drawings: Figure 2 - to $\nu = 0,017$ Hz and Figures 3 and 4 - respectively for 0.17 and 1.7 Hz.

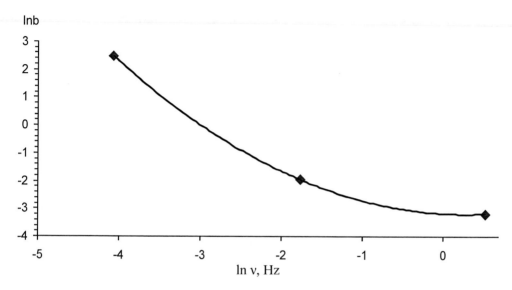

Figure 5. Dependence of parameter "b" from frequency of loading.

In the Figure 2 no jumps in the dependence, in FigIf on fig 3 and 4, the values for the initial periods of exposure time is several times higher than those in the subsequent course of dependencies. Perhaps this difference is caused by various structural transformations under mechanical loading with different frequencies.

With regard to the activation energy, it is the smallest (2.4 kJ / mol) for the frequency of 1.7 Hz and for frequencies of 0.17 and 0.017 Hz, the activation energy, calculated according to the results of these experiments, it turns out the same and equal to 4.9 kJ.

To paraphrase the equation of the Aleksandrov-Gurevich (18), where instead of the stress (σ) using the frequency ν, instead of the coefficient "a" use the "" b, then we can define a certain characteristic value, similar to U_0, the initial activation energy of relaxation process, a constant:

$$U_0 = U + b\nu$$

Analysis of the dependence of the initial activation energy U_0 of the frequency (Figure 6) shows that with increasing frequency ν decreases linearly energy U_0 frequency of loading, Hz

Thus, using the principle of temperature-time superposition and kinetic coefficient of mechanical losses can be at different intensities of loads to determine the time interval in which the measurement of relaxation parameters will be most correct. Besides using the above approximation, we can give a preliminary assessment of the relaxation parameters and analyze the nature of relaxation processes taking measurements without changing the initial temperature.

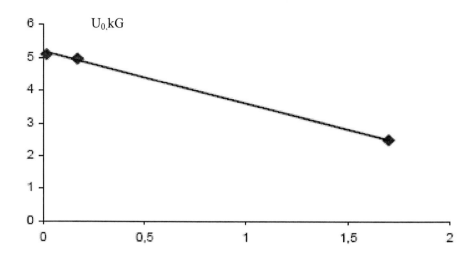

Figure 6. Dependence of initial energy of activation relaxation process from frequency of loading.

REFERENCES

[1]   G. Nikolis, I. Prigogine. *Exploring Complexity. An introduction.* W. H. Freeman and Company//New York.1997.

[2]   Luce, RD, Krantz, DH, Suppes, P., and Tversky, A. Foundations of measurement.(Representation, xiomatization and invariance)//New York: Academic Press. 1998.

[3]   R. B. Turner. Stress relaxation properties of some reaction injection molded (RIM) and reinforced RIM systems./Polymer Composites.-V. 5.- Iss. 2.- p. 151–154. 1984.

[4]   R.M. Bartenev, S.V. Baglyuk and V.V. Tulinova. Microstructure of the polymer chains and relaxational properties of the polybutadienes.//Polymer Science.-V. 32, Iss.7.- 1990.- P. 1364-1371 .

[5]   V. A .Shershnev, S. V. Emeljanov. Reokinetichesky researches of formation of mesh structures in polymers.//Review MITHT.-2006-V.1.- № 5.-P.3-19.

[6]   A .J. Malkin. The Current state of a rheology of polymers: achievements and problems.//Vysokomolek.soedin. - 2009.-V.A 51. № 1. P.106-136.

[7]   A. J. Malkin. Methods of measurement of mechanical properties of polymers.//Moscow.: *Chemistry.* 1978.-336 p.

[8]   A.A.Askadsky, V.M.Markov, O.V.Kovriga, etc. the Analysis of a relaxation of pressure in nonlinear area of mechanical behavior of polymers.//*Vysokomolek. Soed.* - 2009.-V.A 51. № 5. P.576-582.

[9]   S.K. Nechaev, A.R. Khokhlov. Polymer chain elasticity in the presence of a topological obstacle./ Physics Letters A.- V. 126, N 7.P.431-433

[10] G.M.Lukovkin, M.S.Arzhakov, A.E.Salko, S.A.Arzhakov. About the nature of the generalized physicomechanical behavior of polymeric glasses.//Deformation and destruction of material. 2006. - №6. P.18-24.

[11] Yu. Ya. Gotlib, I. A. Torchinskii, V. P. Toshchevikov and V. A. Shevelev . Theory of Relaxation Spectra for Two Identical Interpenetrating Polymer Networks. / *Polymer Science*, A, Vol. 52, No.1, 2010.P.82-93.

[12] V. I. Irzhak Relaxation properties of polymers and the physical network model.// *Russ Chem. Rev. Vol. 69, N. 3.*- 2000.-P. 261-265.

[13] J. Ferry. Viscoelastic properties of polymer/Wiley; 3 edition.1980.- 672 p.

[14] M. Doi and S.F. Edwards, The theory of polymer dynamics, Oxford University Press, Oxford, 1986.

[15] J.K.C. Suen, Y.L. Joo and R.C. Armstrong, Molecular orientation effects in viscoelasticity, *Annu. Rev. Fluid Mech.,* V. 34.-2002. P. 417-444.

[16] V.A.Lomovsky, Z.I.Fomkina, V.L.Bulba, etc. Research Technique of the relaxation phenomena of polymers in a high elastic condition. (Dynamic methods). MITHT.- 2010.-35 p.

In: Polymer Yearbook – 2011.
Editors: G. Zaikov, C. Sirghie et al. pp. 221-230

ISBN 978-1-61209-645-2
© 2011 Nova Science Publishers, Inc.

Chapter 21

THE NEW APPROACH TO RESEARCH OF MECHANISM CATALYSIS WITH NICKEL COMPLEXES IN ALKYLARENS OXIDATION

L. I. Matienko, L. A. Mosolova,
V. I. Binyukov, E. M. Mil and G. E. Zaikov

N. M. Emanuel Institute of Biochemical Physics, Russian Academy of Sciences,[1]
119334 Moscow, Russian Federation

ABSTRACT

AFM the method has been used in the analytical purposes, namely, for research of possibility of formation of the stable supramolecular structures on the basis of binuclear heteroligand complexes Q = Ni2(AcO)3(acac)MP·2H2O (MP = N-methylpirrolidon-2) at the expense of H-bonding. Earlier it has been established by us that complexes Q really are the active particles, formed in the course of alkylarens oxidation to hydroperoxides with molecular oxygen, catalyzed by system {Ni(acac)2 + MP}.

Keywords: catalysis, selective oxidation, alkylarens (ethylbenzene), hydroperoxide, molecular oxygen, $Ni_2(AcO)_3(acac)MP·2H_2O$ (MP = N-methylpirrolidon-2), H-bonding, AFM technique.

I. AIM AND BACKGROUND

The most challenging problem is the selective oxidation of hydrocarbons to hydroperoxides, the primary oxidation products. It should be noted that no efficient catalysts, except for Ni(ll)-based catalytic systems proposed by Matienko et al [1] were found so far for the selective oxidation of ethylbenzene to α-phenylethylhydroperoxide, despite the fact that

---

[1] Kosygin 4, 119334 Moscow, Russian Federation. Fax (7-495) 137 41 01, e-mail: matienko@sky.chph.ras.ru.

the ethylbenzene oxidation catalyzed by both homogeneous and heterogeneous catalysts was sufficiently well studied and described in many publications [2-5].

In addition to academic interest, the problem of selective oxidation of alkylarenes to hydroperoxides is economically sound. Hydroperoxides are used as intermediates in the large-scale production of important monomers. For instance, propylene oxide and styrene are synthesized from α-phenylethylhydroperoxide, and cumyl hydroperoxide is the precursor in the synthesis of phenol and acetone [6].

A method of modifying the Ni(ll) and Fe(II,III) complexes used in the selective oxidation of alkylarenes (ethylbenzene and cumene) with molecular oxygen to afford the corresponding hydroperoxides aimed at increasing their selectivities was first proposed by Matienko [1] consists of introducing additional mono- or multidentate modifying ligands into catalytic metal complexes. The mechanism of action of such modifying ligands was elucidated. New efficient catalysts of selective oxidation of ethylbenzene to α-phenylethylhydroperoxide and cumene to cumyl hydroperoxide on the basis of nickel compounds were developed [1, 7]. However many questions remained while without the answer. So, the structure of the active intermediate complexes formed in situ during oxidation responsible for growth of selectivity and degree of conversion of oxidation of alkylarens to hydroperoxides has been investigated [1, 7]. During too time it was not clear, with what stability of these complexes in relation to the processes leading to their collapse is connected. In the present work with use of AFM method attempt to answer this question is made.

II. INTRODUCTION

The high activity of catalytic systems {M(L1)n + L2} (M=Ni(II), Fe(II,III), L1 =acac− (Ni(II), Fe(II,III), enamac− (Ni(II), L2 = crown-ethers or quaternary ammonium salts, HMPA, N-methylpirrolidon-2 (MP)) as catalysts of ethylbenzene oxidation into α-phenylethylhydroperoxide (PEH) first modeled by us (the increase in SPEH,max , conversion, and yield of PEH may be connected with the stability of catalytic active intermediate heteroligand complexes MIIxL1y(L1ox)z(L2)n (L1ox = CH3COO−) with respect to oxygenation by molecular oxygen into inactive catalytic form. Growth of stability of complexes MIIxL1y(L1ox)z(L2)n to oxygenation,, seems to be due to the formation of intra- and intermolecular H-bonds [7].

Nanostructure science and supramolecular chemistry are fast evolving fields that are concerned with manipulation of materials that have important structural features of nanometer size (1 nm to 1 μm) [8, 9]. Nature has been exploiting noncovalent interactions for the construction of various cell components. For instance, microtubules, ribosomes, mitochondria, and chromosomes use mostly hydrogen bonding in conjunction with covalently formed peptide bonds to form specific structures. The self-assembled systems and self-organized structures mediated by transition metals are considered in connection with increasing research interest in chemical transformations with use of these systems.

The mechanism of catalysis often involves the formation of a supramolecular assembly during the reaction [10].

H-bonding can be a remarkably diverse driving force for the self-assembly and self-organization of materials. H-bonds are commonly used for the fabrication of supramolecular assemblies because they are directional and have a wide range of interactions energies that are tunable by adjusting the number of H-bonds, their relative orientation, and their position in the overall structure. H-bonds in the center of protein helices can be ca. 20 kcal/mol due to cooperative dipolar interactions [11, 12].

Here at first AFM method was used in analytics aims, for research of possibility of the supramolecular structures formation on basis of binuclear heteroligand complex $Ni2(AcO)3(acac)L2·2H2O$ (L2 = MP) at the expense of H-bonding.

III. Experimental

AFM SOLVER P47/SMENA/ with Silicon Cantilevers NSG11S (NT MDT) with curvature radius 10 nm, tip height: $10 - 15$ μm and cone angle $\leq 22°$ in taping mode on resonant frequency 150 KHz was used.

As substrate the polished Silicone chemically modified with trimethylchlorsilan was used $((CH3)3SiOSi)$.

Waterproof $(CH3)3SiOSi$ was exploit for the self-assembly-driven growth due to H-bonding of binuclear heteroligand complex $Ni2(AcO)3(acac)MP·2H2O$ (Q) with $(CH3)3SiOSi$ surface. The saturated solution of complex Q (the complex received in the powdery form (see below)) in water was put on a surface, maintained some time, and then water deleted from a surface by means of special method – spin-coating process.

In the course of scanning of investigated samples it has been found, that the structures are fixed on a surface not strongly enough to spend measurements in contact mode. So samples are well measured in taping mode. At that in additional experiments it has been shown that at scanning there is a turning movement and orientation of investigated structures with cantilever.

The received results are presented on the Figures 1-5 (Part IV).

IV. Results and Discussion

The method of transition metal catalysts modification by additives of electron-donor ligands for increase in selectivity of liquid-phase alkylarens oxidations into corresponding hydroperoxides was proposed by us [1]. It was established that the high activity of binary systems {ML1n + L2} (M=Ni, Fe, L1 are acetylacetonate (acac) or enaminoacetonate (enamac) ions, L2 are monodentate ligands (MP, HMPA) or crown ethers and quaternary ammonium salts) is associated with the fact that during the alkylarens oxidation, the active primary complexes (MIIL12)x(L2)y (I macro step of oxidation) and heteroligand complexes MIIxL1y(L1ox)z(L2)n (L1ox is (OAc)– ion) (II macro step of oxidation) are formed to be involved in the oxidation process. The third macro step of alkylarens oxidation is connected with catalytic action of M(OAc)2. The latter complex is the final transformation product of the primary complex (MIIL12)x(L2)y.

It was shown [1] that the selective catalyst Ni(acac)(AcO)·L2 was formed as a result of the ligand L2-controlled regioselective addition of O2 to a nucleophilic carbon atom (γ-C) of the acac ligand. The coordination of the electron-donating axial ligand L2 with the Ni(acac)2 complex stabilized the intermediate zwitter-ion [L2(acac)Ni(acac)+ ⋯ O2−] and increased the probability of regioselective insertion of O2 to the acetylacetonate ligand activated by coordination with the nickel(II) ion. Further incorporation of O2 into the chelate acac-ring was accompanied by the proton transfer and the redistribution of bonds in the transition complex A leading to the scission of the cyclic system (A) to form a chelate ligand OAc−, acetaldehyde and CO (in the Criegee rearrangement) (Scheme 1).

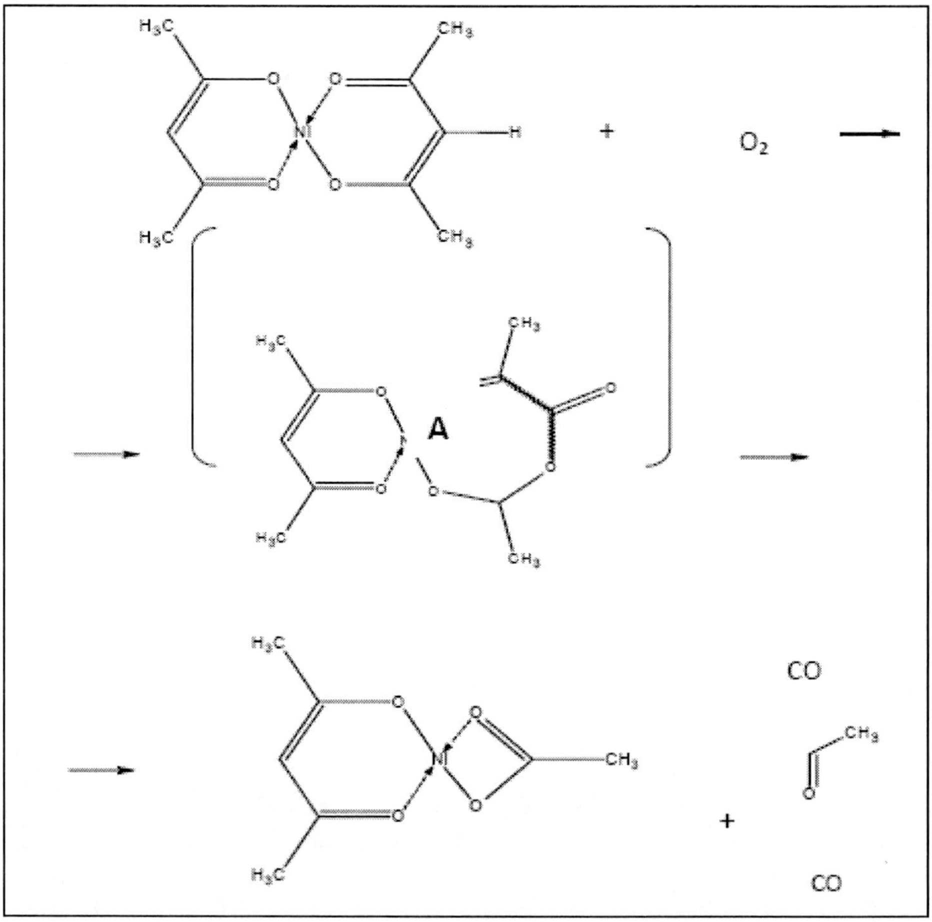

Scheme 1.

As a result of this process, reactive mono- and poly-nuclear heteroligand complexes with the general composition Nix(acac)y(AcO)z(L2)n were formed [1]. It is known that heteroligand complexes are more active in relation to reactions with electrophiles in comparison with homoligand complexes [7]. Thus the stability of heteroligand complexes Nix(acac)y(AcO)z(L2)n with respect to conversion into inactive catalytic form seems to be due to the formation of intermolecular H-bonds.

Ni(acac)2 L2 + O2 ⟶ [L2(acac)Ni(acac)+ ⋯ O2−] ⟶ L2(acac)Ni(AcO) + MeCHO + CO,

2Ni(acac)2 L2 + O2 ⟶ Ni2(AcO)3(acac)L2 + 3MeCHO + 3CO + L2.

(L2 = HMPA, DMF, MP, M'St (M' = Na, Li, K)).

The only known Ni(II)-containing dioxygenase ARD (acireducton dioxygenase), which catalyzes the oxidative decomposition of β-diketones, operates in the analogous way [13]. This applies to the functional enzyme models, namely, Cu(II)- and Fe(II)-containing quercetin 2,3-dioxygenases, which catalyze the decomposition of β-diketones in the enol form to carbonyl compounds with CO evolution [14, 15].

The complex, formed in the course of ethylbenzene oxidation, catalyzed with system {Ni(acac)2 + MP}, has been synthesized by us and its structure has been defined with mass spectrometry, electron and IR spectroscopy and element analysis [1]. The certain structure of a complex Ni2(AcO)3(acac)MP·2H2O ("Q") corresponds predicted on the basis of the kinetic data [1].

It is proved that the complex "Q" catalyzes selective ethylbenzene oxidation into α-phenylethylhydroperoxide (Figure1, curve 1). Besides it has been established that the product "Q" is a really binuclear heteroligand complex of nickel, instead of is mechanical mixture Ni(acac)(OAc)·MP with nickel acetate. It follows from the following data: it has been shown that in the presence of Ni(OAc)2 ethylbenzene is oxidized with formation of phenol as the main product: selectivity of process on PEH SPEH = 70-80 % only at depths of oxidation not above C =1-2 %, and then sharply falls (Figure1, curve 3).

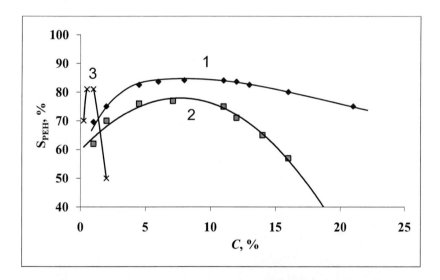

Figure 1. Dependences SPEH on C in the ethylbenzene oxidation: at catalysis with Ni2(AcO)3 (acac)MP·2H2O (1), system {Ni(acac)2+MP} (2), Ni(OAc)2 (3). [Ni(acac)2]=[Ni(OAc)2]=1·10-3, [MP]= 1·10-2, ["Q"]=5·10-4 mol/l, 1200C.

Prospective structure of the complex Ni2(AcO)3(acac)MP·2H2O "Q" is presented with Scheme 2.

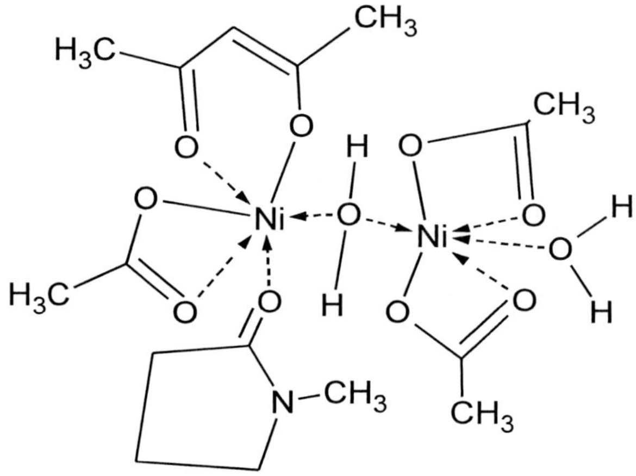

Scheme 2.

FORMATIONS OF MACRO STRUCTURES AT THE EXPENSE OF H-BONDING ON ((CH3)3SIOSI SURFACE. (THE USE OF AFM METHOD)

On the basis of the known from the literature facts it was possible to assume that binuclear heteroligand complexes Ni2(OAc)3(acac)·MP·2H2O are capable to formation of macro structures at the expense of intermolecular H-bonds (H2O – MP, H2O – acetate (or acac–) group) [16, 17].

The association of Ni2(AcO)3(acac)L2·2H2O to supramolecular structures as result of H-bonding is demonstrated on the next Figures (2-6).

On Figure 2–6 three-dimensional and two-dimensional AFM images (10×10(μm), 2×2(μm) and 0.35×0.35(μm)) of the structures formed at drawing of a uterine solution on a surface of modified silicone are presented (μm = micron (micrometer) = 103 nm). It is visible that the majority of the generated structures have rather similar form of three almost merged spheres.

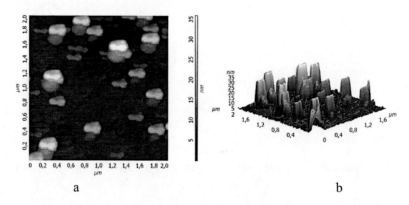

a   b

Figure2. The AFM two- (A) and three-dimensional (B) images of nanoparticles on the basis Ni2(AcO)3(acac)L2·2H2O formed on the surface of modified silicone.

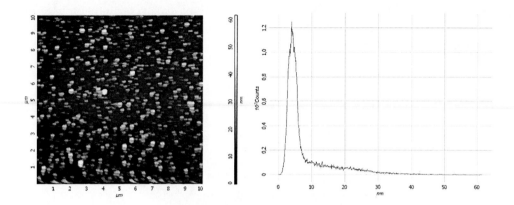

Figure 3. AFM image of structures on the modified silicone surface 10×10 μm (at the left).The distribution histogram on height of nanoparticles (to the right).

As seen from the distribution histogram on Figure 3, the greatest number of particles - is particles of the size 3-4 nm on height. As it is possible to see in (Figure2-A), except particles with the form reminding three almost merged spheres, there are also structures of more simple form (with the height approximately equal 3-4 nm).

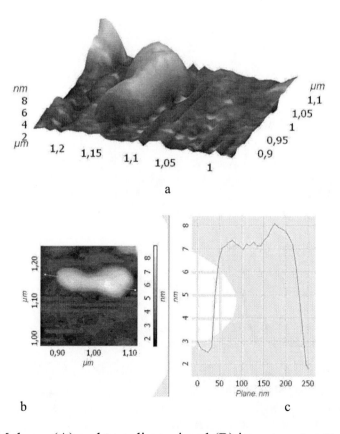

Figure 4. The AFM three- (A) and two-dimensional (B) images and profile of the structure (C) with minimum height along the greatest size in plane XY.

On Figure 5 three-dimensional image of one of structures having the form of three almost merged spheres (two - relatives on the size, and one - it is appreciable the smaller size) is presented.

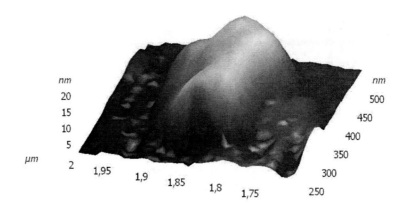

Figure 5. Three-dimensional image of one of structures having the form of three almost merged spheres.

On Figure 6 profiles of this structure to planes XY in three various directions are presented.

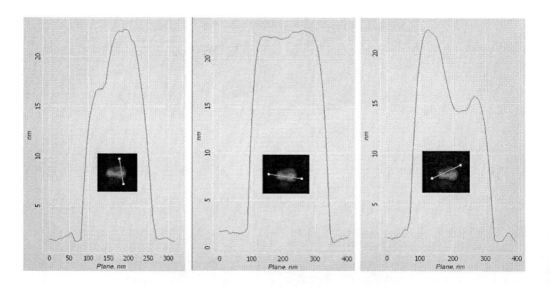

Figure 6. Two-dimensional image of one of structures having the form of three almost merged spheres and profiles in three various directions in plane XY.

From the received data the following is visible. It is important to notice that for all structures the sizes in plane XY do not depend on height on Z (Figure 4, 6). They make about 200 nm along a shaft which are passing through two big spheres, and about 150 nm along a shaft crossing the big and smaller spheres (Figure 6). But all structures are various on heights from the minimal 3-4 nm to ~ 20-25 nm for maximal values. This is visible on profiles of some images (Figure 4, 6), and is in the good consent with the histogram data (Figure 3).

It follows from this that structure with the minimum height make the most part of their total. In distribution on height there is a small quantity of particles with maximum height 20-25 nm and considerably smaller quantity with height to 35 nm (Figure 2, 6).

Thus, in the present article we have shown what the self-assembly-driven growth seems to be due to H-bonding of binuclear heteroligand complex Ni2(AcO)3(acac)MP·2H2O ("Q") with (CH3)3SiOSi surface, and further due to directional inter-molecular H-bonds, apparently at parcipitation of H2O molecules, acac, acetate groups, MP [16, 17] (see Scheme 2).

CONCLUSION

Earlier [7] the important role of H-bonds in the formation of active catalytic species, formed during alkylarens (ethylbenzene) oxidation in the presence of {ML1n + L2} (Scheme 1), was confirmed by us at studying the effect of addition of small amounts of water to catalytic systems {ML1n + L2}. It was found that the introduction of small portions of water to catalytic systems based on iron complexes leads to the higher catalytic activity of these systems.

Here AFM method was used by us in the analytical purposes to research the possibility of the formation of supramolecular structures on basis of binuclear heteroligand complex Ni2(AcO)3(acac)L2·2H2O (L2 = MP) at the expense of H-bonding.

The experimental data, presented in Figure 2-6, are not the demonstration of formation of similar macro structures in the real conditions of Ni-complexes catalyzed alkylarens oxidation. These data specifies in quite probable possibility of formation supramolecular structures on the basis of complex Q at the expense of H-bonds formation. H-bonding seems to be one of the factors, responsible for the stability of heteroligand complexes "Q", the intermediate products of oxygenation of primary complexes Ni(acac)2·MP, and therefore the high values of selectivity, and conversion of alkylarens (ethylbenzene, cumene) oxidation into hydroperoxide in the developed reaction of oxidation at the presence of system {Ni(acac)2 + MP}.

REFERENCES

[1] L.I. Matienko, Solution of the problem of selective oxidation of alkylarenes by molecular oxygen to corresponding hydroperoxides. Catalysis initiated by Ni(II), Co(II), and Fe(III) complexes activated by additives of electron-donor mono- or multidentate extra-ligands, In: Reactions and Properties of Monomers and Polymers, Ed. by A. D'Amore and G. Zaikov. New York: Nova Sience Publ. Inc., 2007, p.21.

[2] N.M. Emanuel, D. Gal, Ethylbenzene oxidation. Model reaction, Moscow: Nauka (1984) (in Russian).

[3] Yu.D. Norikov, E.A. Blyumberg, L.V. Salukvadze, Role of a surface of the heterogeneous catalyst in the mechanism of liquid-phase hydrocarbons oxidations, In: Problemy kinetiki i kataliza, Moscow: *Nauka,* 16, 150 (1975) (in Russian).

[4]   M.V. Nesterov, V.A. Ivanov, V.M. Potekhin, V.A. Proskuryakov, M.Yu. Lysukhin, Catalytic activity of the mixed catalysts at ethylbenzene oxidation, Zh. Prikl. Khimii, 52, 1585 (1979) (in Russian).

[5]   P.P. Toribio, J.M. Campos-Martin, J.L.G. Fierro, Liquid-phase ethylbenzene oxidation in hydrocarbon with barium catalysts, *J. Mol. Catal. A: Chem.,* 227, 101 (2005).

[6]   K. Weissermel, H.-J. Arpe, Industrial Organic Chemistry, 3nd ed., transl. by Lindley C.R.. New York: VCH (1997).

[7]   L.I. Matienko, L.A. Mosolova, G.E. Zaikov, *Selective Catalytic Hydrocarbons Oxidation. New Perspectives,* New York: Nova Science Publishers, Inc. 2010, 158 P.

[8]   St. Leninger, B. Olenyuk, P.J. Stang, Self-Assembly of Discrete Cyclic Nanostructures Mediated by Transition Metals, *Chem. Rev.,* 100, 853 (2000)

[9]   P.J. Stang, B. Olenyuk, Self-Assembly, Symmetry, and Molecular Architecture: Coordination as the Motif in the Rational Design of Supramolecular Metallacyclic Polygons and Polyhedra, *Acc. Chem. Res.,* 30,502 (1997)

[10]   Beletskaya, V.S. Tyurin, A.Yu. Tsivadze, R. Guilard, Ch. Stem, Supramolecular Chemistry of Metalloporphyrins, *Chem. Rev.,* 109, 1659 (2009)

[11]   C.M. Drain, A. Varotto, I. Radivojevic, Self-Organized Porphyrinic Materials, *Chem. Rev.,* 109, 1630 (2009)

[12]   Cheng-Che Chu, G. Raffy, D. Ray, A. Del Guerzo, B. Kauffmann, G. Wantz, L. Hirsch and D.M. Bassani, Self-Assembly of Supramolecular Fullerene Ribbons via Hydrogen-Bonding Interactions and Their Impact on Fullerene Electronic Interactions and Charge Carrier Mobility, *J. Am. Chem. Soc., 132,* 12717 (2010)

[13]   Y. Dai Y., Th. C. Pochapsky, R.H. Abeles, Mechanistic Studies of Two Dioxygenases in the Methionine Salvage Pathway of Klebsiella pneumoniae, *Biochemistry,* 40, 6379 (2001)

[14]   .B. Gopal., L.L. Madan, S.F. Betz, and A.A. Kossiakoff, The Crystal Structure of a Quercetin 2,3-Dioxygenase from Bacillus subtilis Suggests Modulation of Enzyme Activity by a Change in the Metal Ion at the Active Site(s), *Biochemistry,* 44, 193 (2005)

[15]   É. Balogh-Hergovich, J. Kaizer, G. Speier, Kinetics and mechanism of the Cu(I) and Cu(II) flavonolate-catalyzed oxygenation of flavonols, Functional quercetin 2,3-dioxygenase models, *J. Mol. Catal. A: Chem.,* 159, 215 (2000).

[16]   E.V. Basiuk, V.V. Basiuk, J. Gomez-Lara, R.A. Toscano, A bridged high-spin complex bis-[Ni(II)(rac-5,5,7,12,12,14-hexamethyl-1,4,8,11-tetraazacyclotetradecane)]-2,5-pyridinedicarboxylate diperchlorate monohydrate, J. Incl. Phenom. Macrocycl. Chem. 38, 45 (2000).

[17]   P.Mukherjee, M.G.B. Drew, C.J. Gómez-Garcia, A. Ghosh, (Ni2), (Ni3), and (Ni2 + Ni3): A Unique Example of Isolated and Cocrystallized Ni2 and Ni3 Complexes, Inorg. Chem., 48, 4817 (2009).

In: Polymer Yearbook – 2011. ISBN 978-1-61209-645-2
Editors: G. Zaikov, C. Sirghie et al. pp. 231-233 © 2011 Nova Science Publishers, Inc.

Chapter 22

5TH INTERNATIONAL CONFERENCE ON "TIMES OF POLYMER (TOP) AND COMPOSITES"

G. E. Zaikov, M. I. Artsis and L. L. Madyuskina

N.M. Emanuel Institute of Biochemical Physics
Russian Academy of Sciences[1]
Moscow 119334, Russia

5th International Conference on "Times of Polymer (TOP) and Composites" was held on June, 20 – 23 2010 on Ischia Island in Hotel "Continental Terme" (Naples Bay), Italy.

This conference was organized by Department of Aerospace and Mechanical Engineering, Second University of Naples – SUN, Department Materials and Production Engineering and University of Naples Federico II.

Prof. Domenico Acierno (Department of Materials and Production Engineering University of Naples Federico II) and Prof. Alberto D'Amore (Engineering Schools of II University of Naples – SUN Department of Aerospace and Mechanical Engineering) were the Co-Chairmen of conference.

World well known scientists were included in Scientific Committee. Domenico Acierno (University of Naples, Italy), Alberto D'Amore (II University of Naples SUN, Italy), David Kranbuhel (College of William and Mary, USA), Gregory B. McKenna (Texas Tech University, USA), Jovan Mijovic (Polytechnic University, Brooklin, USA), Luigi Nicolais (University of Naples, Italy), George Papanicolaou (University of Patras, Greece), Sindee Simon (Texas Tech University, Usa), Graham Williams (University of Wales Swansea, U.K.), Guennadi E. Zaikov (Inst. of Bioch. Phy. Moscow, Russia), Carla Minarini (ENEA-Italy), Francesco Ciardelli (University of Pisa, Italy), Jane Lipson (Dartmouth College, USA), Anne Hiltner (Oregon State University, USA), Jean Luc Gardette (CNRS, France), Igor Emri (University of Ljubljana, Slovenia) were members of Scientific Committee of conference.

The conference provided a forum for scientists and engineers throughout the world interested in the timescales of polymers and composites processing, structure and properties.

[1] 4, Kosygin st., Moscow 119334, Russia,chembio@sky.chph.ras.ru.

As time is the driving concept in the polymer science community, TOP-Conferences included sessions on various topics and provided opportunities for exchanging ideas and opinions on both fundamental science and industry-relevant subjects.

TOP was an evolving, dynamic conference that embraces cutting edge research topics and emerging scientists. It was thought having as a primary objective the meeting of a number of scientists working within the area of timescales of polymers, conceived as the background driving force for the progress of knowledge in many field of Polymer Science.

The conference program focused on the more recent advances in the following topics:

- Viscoelasticity/Rheology
- Glass Transition
- Adhesion
- Processing
- Durability/Degradation
- Biomaterials
- Fracture/Yielding
- Sensors
- Thin Films
- Composites/Nanocomposites
- New Techniques
- Transport phenomena

The program of the conference included 1 plenary lecture, 20 invited lectures, 64 oral presentations and poster sessions.

Some oral sessions were included in scientific program. The chairmen of these sessions were Kia L. Ngai, Sindee Simon, David Kranbuehl, Ming Qiu Zhang, Gregory B. McKenna, Gennady Zaikov, Mike Roland, Richard Wool, Marina G. Guenza.

Plenary lecture was done by Prof. G. B. McKenna (Texas Tech University, USA) and was devoted to interrogating the physics of materials: mechanics of materials from glass to rubber and from the macro to the nano.

20 invited lectures were included in program. M.Q. Zhang (Zhongshan University, P. R. China) gave presentation about self-healing polymers and composites, preparation and characterization and D. Papaspyrides (National Technical University of Athens) spoke about nanotechnology and food contact materials

Tailoring the structure and dynamics of polymer/brush-coated nanoparticle systems was discussed in invited lecture of P. F. Green (University of Michigan, USA) and swelling the molecular entanglement network in polymer glasses were presented by K. Dalnoki-Veress (McMaster University, Canada).

M. Roland (Naval Research Laboratory, USA) gave information about prediction of elastomer service lifetimes and J. Lipson (Dartmouth College, USA) spoke about predicting glass transitions in thin film polymers.

The next two invited lectures were devoted to problems elastic and viscoelastic behavior of polymer matrix MWCNT nanocomposites (G. Papanicolau, University of Patras, Greece) and Intermolecular effects in the dynamics of polymer melts: interplay of cooperative dynamics and entanglements (M. G. Guenza, University of Oregon, USA). The title of the lecture of V. Kulichkin (Russian Academy of Sciences, Russia) was "From rheology of nanocomposites to rheology of polymer melts: step back or forward?"

The talk of K. Friederich (University of Kaiserslautern, Germany) had a title "On sliding wear of nanoparticle modified polymer composites" and J. Seferis (USA) spoke about nano free volume : a concept for blurring the solid liquid and gaseous states in polymeric composites.

The invited lecture of A. Hiltner (Case Western Reserve University, USA) was devoted to confined crystallization of polymers in coextruded nanolayer assemblies and reaction in the melt of post-consumer poly(ethylene terephthalate) (PET) with ester functionalized polyolefins was discussed in the invited talk of F. Ciardelli (University of Pisa, Italy).

J.-L. Gardette (CNRS, France) spoke about predicting the ageing and the long-term durability of organic polymer solar cells and I. Emri (University of Ljubljana, Slovenia) gave presentation under the title "On the behaviour of dynamically loaded polymeric".

The viscoelastic bulk modulus, effect of macromolecular structure was discussed in invited lecture of S. Simon (Texas Tech University, USA) and creating a uniform dispersion of surface functionalized graphene nanosheets in polymers, characterizing the polymer-particle interface and mechanical properties were presented in the invited talk of D. Kranbuhel (College of William and Mary, USA).

The titles of the two last invited lectures were "On the universal properties of relaxation and diffusion in complex interacting systems" (K. L. Ngai, Naval Research Laboratory, USA) and "Elasticity and inelasticity of hard-phase reinforced polyurethane elastomers: from sensitivity to chemical and physical structure to time dependent phenomena" (C. Prisacariu, Institute of Macromolecular Chemistry Petru Poni, Romania).

The lectures which were included in some sessions of the conference were devoted to the next problems: using polarized neutrons for elastic and dynamic studies on protein systems; nucleation of polyethylene crystals; twinkling fractal theory of the glass transition and yield stress; Au based nanocomposites towards plasmonic applications; comparing nanofillers in polylactide nanocomposites; design of novel polymeric materials by controlled selfassembly; investigation of model fuel effects on thermal oxidation of polyethylene; mass transport in nanocomposite materials for membrane separations; predicting the photoageing and photostabilization of polymer nanocomposites; effect of the compounding procedure on the structure and viscoelasticity of polymer nanocomposites.

Poster session included 60 presentations. Particularly there was the poster of Prof. G. Zaikov (Institute of Biochemical Physics, Russia) "Thermal degradation and combustion behavior of polypropylene/MWCNT composites"

The conference was shown that synthesis, properties and application of polymers and polymer composites are very important things for pure and applied chemistry and for material science first of all.

The next 6th conference will be held in the same place in two years (June, 2012).

In: Polymer Yearbook – 2011.
Editors: G. Zaikov, C. Sirghie et al. pp. 235-238

ISBN 978-1-61209-645-2
© 2011 Nova Science Publishers, Inc.

Chapter 23

14TH INTERNATIONAL SCIENTIFIC CONFERENCE ON "POLYMERIC MATERIALS"

G. E. Zaikov, M. I. Artsis and L. L. Madyuskina

N.M. Emanuel Institute of Biochemical Physics
Russian Academy of Sciences[1], Moscow 119334, Russia

14th International Scientific Conference on "Polymeric Materials" was held on September, 15 – 17 2010 in Halle (Saale), Germany. This conference was organized by Martin Luther University Halle-Wittenberg and Polymer Competence Center Halle-Merseburg in cooperation with Innovation Center Polymer Technology.

Prof. Hans-Joachim Radusch (Polymer Competence Center Halle-Merseburg) was the conference chairman. World well known scientists were included in Program Committee: René Androsch, Michael Bartke, Mario Beiner, Wolfgang Grellmann, Thomas Groth, Jörg Kreßler, Goerg Michler, Wolfgang Paul, Thomas Thurn-Albrecht, Ralf Wehrspohn.

About 350 participants from 100 research centers of 20 countries (Germany, Russia, Vietnam, Japan, France, Georgia, Czech Republic, Romania, Uzbekistan, Iran, Poland, Austria, USA, Hungary, South Africa, Algeria, South Korea, Ukraine) took part in this conference.

"Polymeric Materials 2010" was extended, thus in plenary, keynote and short lectures as well as poster presentations scientific results in the field of

- Polymer Chemistry
- Polymer Physics
- Polymer Engineering
- Polymers in Energy Engineering

were presented and discussed.

The program of the conference included 4 plenary lectures, 17 keynote lectures, short presentations and poster sessions.

---

[1] 4, Kosygin st., Moscow 119334, Russia, chembio@sky.chph.ras.ru.

Fundamentals of fine particle mixing in polymers were discussed in the first plenary lectures (Manas-Zloczower I., Cleveland, USA). The second plenary lecture was about actively moving polymers (Lendlein A.; Behl M., Teltow and Berlin, Germany). The topic of the third plenary lecture was "Macromolecules, assemblies, particles – a discovery journey in materials synthesis (Klapper M.; Weil T.; Mullen K., Mainz, Germany). The last plenary lectures was devoted to making polymers swim (Mykhaylyk O.O.; Ryan A., Sheffield, UK)

Symposium "Polymer Chemistry" focused on the following topics:

- Synthesis and Characterization
- Biobased Polymers
- Polymers in Biomedical Application
- Modification and Interfaces

This symposium included 4 keynote lectures and 20 short lectures. Barner-Kowollik C.; Inglis A.J.; Nebhani L.; Blinco J.; Paulohrl T.; Glassner M. (Karlsruhe, Germany) gave keynote lecture about constructing functional materials via (reversible) diels-alder chemistry. Internal composition and mechanical properties of polyelectrolyte multilayer films containingpolysaccharides were discussed in the keynote lecture of Picart C.; Boudou T.and Crouzier T. (Grenoble and Montpellier, France). Fink H.-P.; Ganster J. and Engelmann G. (Potsdam-Golm, Germany) spoke about biobased polymers and composites. The last keynote lecture of this symposium was devoted to the influence of stearic acid coating on the viscosity of hydrated filler suspensions (Focke W.W.; Molefe D.; Labuschagne F.J.W.; Ramjee S., Pretoria, South Africa)

The short lectures of this symposium were devoted to the next problems: polymer material design via topological control using RAFT polymerization; effects of electric fields on block copolymer nanostructures; block copolymer nanotubes by melt-infiltration of nanoporous aluminum oxide; multifunctional dendritic polyglycerol architectures for drug and dye delivery; protein folding and misfolding studied by NMR spectroscopy; synthesis and properties of triphilic polymers; HB polymers in coatings and thin film application.

Symposium "Polymer Physics" included topics:

- Theory and Modelling
- Polymer Dynamics
- Novel Experimental Techniques
- Structure and Morphology

4 keynote and 26 short lectures were done during this symposium. The title of the first keynote lecture was "60 years after Flory's ideality hypothesis: Are polymer chains in a melt really ideal?" (Meyer H.; Wittmer J.P.; Johner A.; Obukhov S.P.; Farago J.; Baschnagel J., Strasbourg, France; Gainesville, USA). Lequeux F. (Paris, France) spoke how filled elastomer mechanical properties are controled by the glassy dynamics around the nanoparticles. Polymer crystallization studied at processing

relevant scanning rates (<106 K/s) was discussed in the keynote lecture of Schick C.and Zhuravlev E. (Rostock, Germany). The last keynote lecture of this symposium was about evaluation of network structure and physical properties of Tetra-PEG gel (Sakai T.; Agaki Y.; Matsunaga T.; Tsutsui Y.;Shibayama M.; Chung U., Tokyo, Japan)

The short lectures of this symposium were devoted to the next problems: swelling behavior of a new class of biohybrid networks; characterization of gas sorption in glassy polymers using experimental and molecular modeling techniques; reversible structuring of photosensitive polymer films by surface plasmons; chain dynamics of polymers confined to ordered nanoporous alumina membranes; distribution of relaxation times in a polymer melt and in miscible polymer blends close to the glass transition; atomic force microscopy studies on the morphology of polymeric thermotropic glazings for overheating protection applications; new methods for monitoring structure evolution in composite fibers by X-ray scattering; direct imaging of nanoscale deformation processes in elastomeric polypropylene; study on micostructure development of PP/PET blend nanocomposite fibers.

The topics of Symposium "Polymer Engineering" were:

- Polymer Processing and Rheology
- Advanced Polymer Materials andApplications
- Polymer Blends and Nanocomposites
- Polymer Properties, Testing and Characterization

This symposium included 5 keynote and 37 short lectures. The title of the first keynote lecture was "Advanced performance of single screw extruders: High speed and alternative extrusion concepts" (Wortberg J.; Gorczyca P.; Gromann M., Duisburg, Germany). Orth P. (Frankfurt, Germany) spoke about plastics and climate protection. Thermal degradation and combustion behavior of polypropylene / multiwalled carbon nanotube composites were discussed in the keynote lecture of Zaikov G.E.; Rakhimkulov A.D.; Lomakin S.M.; Dubnikova I.L.; Shchegolikhin A.N.; Davidov E.Y. and Kozlowski R. (Moscow, Russia; Poznan, Poland). The topic of the next keynote lecture was "New insights into polymer-particle nanocomposites: From nano-scale polymer-mediated interactions to macro-scale structural organization" (Heinrich G.; Chervanyov, A.; Saphiannikova M.; Richter S.; Stockelhuber K.W., Dresden, Germany). The last keynote lecture was about fatigue behavior of polymer blends and

Nanocomposites (Altstadt V.; Fischer F.; Gotz C.; Wolff-Fabris F., Bayreuth, Germany)

The short lectures of this symposium were devoted the next problems: using pressure dependent viscosity for improving the temperature calculation of single screw extruders; calender lines inline-control of materials data by process viscometer; viscoelastic behavior of polymer based electrorheological fluids; physical foaming of advanced polymer materials; analysis and modeling of the heating and sealing behavior of polymer films during ultrasonic sealing; modification of reinforcing materials by plasma; filler design for layered double hydroxide nanocomposites; influence of halogen free flame retardant additives on compact and expanded polyester material; influences of polymer matrix melt viscosity and molecular weight on MWNT agglomerate dispersion; time dependent reinforcement effect of nanoclay in rubber composites; a new fatigue test machine for accurate crack growth analysis in rubber compounds; gradients in composition, morphology and properties of semicrystalline polymers.

Symposium "Polymer in Energy Engineering" focused on the topics:

- Polymers for Energy Transportation
- Polymers for Energy Storage
- Polymers for Regenerative Energy Production
- Polymers for Advanced Batteries

This symposium included 3 keynote and 7 short lectures. Lang R.W.; Wallner G.M. (Linz, Austria) gave the keynote lecture about polymeric materials for solar energy application. Novel (co)polymers containing the fluorene building block were discussed by Scherf U. (Wuppertal, Germany). The topic of the last keynote lecture of this symposium was "Block copolymers beyond templating – merging structure and function" (Gutmann J.S., Mainz, Germany).

The short lectures of this symposium were devoted the next problems: correlation of electrical conductivity and mechanical properties of electroconductive composites based on elastomeric or thermoplastic matrices; polymers as solar cell encapsulate materials for application in photovoltaic modules; aging behavior of polymeric absorber materials for solar thermal collectors; investigations of structural changes by modifications of polymer systems for photovoltaic applications; development and characterization of advanced silicon based thermoplastic elastomers for PV encapsulation; side-chain sulfonated poly(ether sulfone)s for PEMFC applications.

Poster session included 153 presentations. Scientists from Russia presented 20 posters. Some of these posters were devoted to the next problems: the degradation heterochain polymers in the presence of phosphorus stabilizers (Kalugina E.V.; Gaevoy N.V.; Gumargalieva K.Z.; Zaikov G.E., Moscow, Russia); mechanism of stable radical generation in lignin under the action of nitrogen dioxide (Davydov E.Y.; Gaponova I.S.; Lomakin S.M.; Pariiskii G.B.; Pokholok T.V.; Zaikov G.E., Moscow, Russia); organosilicon polymers with photo switchable fragments in the side chain (Doroshenko M.; Koynov K.; Zaikov G.E.; Mukbaniani O.V., Tbilissi, Georgia; Mainz, Germany; Moscow, Russia); metallocomplex catalysis in selective alkylarenes oxidations to hydroperoxides, design of new effective Ni (Fe) – catalytic systems, the important role of H-bonds in the mechanism of catalysis (Matienko L.I.; Mosolova L.A.; Zaikov G.E., Moscow, Russia); electric conductivity of polymer composites at mechanical relaxation (Aneli J.N.; Mukbaniani O.V.; Zaikov G.E.; Markarashvili E.G., Tbilissi, Georgia; Moscow, Russia); biodegradation and medical application of microbial poly(3-hydroxybutyrate) (Artsis M.I.; Bonartsev A.P.; Iordanskii A.L.; Bonartseva G.A.; Zaikov G.E., Moscow, Russia).

This conference was shown that topics which were discussed during of the meeting are very important for pure and applied chemistry (research, development, production).

The next conference will be held in 2012 again in Halle-Saale.

In: Polymer Yearbook – 2011.
Editors: G. Zaikov, C. Sirghie et al. pp. 239-241

ISBN 978-1-61209-645-2
© 2011 Nova Science Publishers, Inc.

Chapter 24

INTERNATIONAL SYMPOSIUM
"RESEARCH AND EDUCATION IN INNOVATION ERA"

G. E. Zaikov, L. L. Madyuskina and R. M. Kozlowski[1]
N.M. Emanuel Institute of Biochemical Physics Russian Academy of Sciences[1],
Moscow 119334, Russia
[1]FAO ESCORENA,[2]Poznan 60-630, Poland

International symposium "Research and Education in Innovation Era" was held on the period November, 10 – 12 2010 in Arad (Romania) on the base of "Aurel Valicu" University (AVU). The organizers of the conference were The Ministry of Education, Research, Youth and Sport of Romania (Bucharest) and AVU (Arad).

About 200 scientists and students from 25 research centers of Romania, Poland, The Netherlands, Russia, Italy, Portugal, Germany, UK, Croatia, Serbia, USA, Spain and Switherland took part in this symposium.

The Organizing Committee included 34 world well-known scientists from these countries. The program of the conference included plenary lectures, parallel sessions and poster session. Rector of AVU Prof. Lizica Mihut took part in opening ceremony of the conference. Two plenary lectures were done by Prof. Rodica Zafiu, University of Bucharest ("Present-Day Tendencies in the Romanian Language") and Dr. Patricia Davies, The European Association for University Lifelong Learning, Director of European Dolceta Project ("Dolceta European Project - education for responsible and sustainable consumers").

Programm included 8 parallel sessions.

"Chemistry and Application Field" session had 4 sections. Section one ("Novel Biochemical Methods and Their Applications") included keynote lecture ("Mass spectrometric approaches for elucidation of "misfolding" and aggregation structures of neurodegenerative proteins: ion mobility-MS and affinity-MS" Michael Przybylski, Laboratory of Analytical Chemistry and Biopolymer Structure Analysis, Department of Chemistry, University of Konstanz) and 5 oral presentations: orbital ion trap-a new vision in

---

[1] 4 Kosygin str., Moscow 119334, Russia, chembio@sky.chph.ras.ru.

[2] 71B, Wojska Polskiego str., Poznan 60-630, Poland, E-mail: Ryszard.Kozlowski@escorena.net.

high resolution mass spectromet, ganglioside structure and composition in brain development and malignant alteration, dementia imaging, identification of epimerization and sulfation pattern in brain chondroitin/dermatan sulfate glycosaminoglycans by advanced mass spectrometry, mass spectrometric determination of ganglioside biomarkers in human anencephaly.

The second section "Trends in Biotechnology" of the first session included four keynote lectures: "Technology transfer-an important tool of effective cooperation" (Ryszard Kozlowski, Institute of Natural Fibers, Poznan, Poland); "Kinetics for the chemistry, biology, medicine and agriculture" (Gennady Efremovich Zaikov, N.M.Emanuel Institute of Biochemical Physics, Russian Academy of Sciences, Moscow Russia); "Novel insect repellent textiles" (Vincent Nierstraz, University of Ghent, Belgium); "Statistical analysis of the microscopic images as a way for investigation of chemical and physical processes in polymers" (Sergei Bronnikov, Russian Academy of Science, Institute of Macromolecular Compounds, Sankt Petersburg, Russia) and two oral presentations: synthesis and anti-tubercular activity of the new isoniazid derivatives with the thioamides, cyclodextrin behavior in magnetic field.

The third section "Environmental Protection" included 3 oral presentations: aspects of monitoring emissions of greenhouse gases generated by anthropic activities, the impact of the oil slime over the surface and underground waters in the district of Gorj, the Mures Floodplain natural park, cost-benefit analysis. The fourth section "Food Engineering and Food Safety" had 3 oral presentations: Romanian food safety and strategy, antioxidant activities and active principles of cinnamon (Cinnamomum Aromaticum Nees) and ginger (Zingiber Officinale Roscoe), studies and experimental research on some plants of the Fabaceae Family of B vitamins content under the action of synthetic fertilizers.

Ten posters were included in the first session of symposium.

The second session "Mechanical, Electrical and Textile Engineering" had 7 sections. First section (plenary section) included 4 lectures: "La gestion de activos electromecanicos ferroviarios: adaptacion de herramienta informatica, modelado y analisis de sensibilidad de costes lcc en funcion de la fiabilidad" (Manuel Pocino Pasias, Fernando Pascual Andreu, Luis Lezaun Martinez de Ubago, Emilio Larrode Pellicer, Universidad de Zaragoza. Espana); "About the Possibility of Optimization and/or Replacing a Thermal Treatment, by Using a Concentrated Energy Source During Machining" (Corina Bokor, Sorin M. Itu, Claudiu Isarie, "Lucian Blaga" University of Sibiu); "DMG – Tehnologii modern de prelucrare pe masinile unelte cu CNC" (Lilian Cirstea, S.C. DMG S.R.L., Arad); "A New Adaptive Teaching Method for Engineering School" (Dorin Isoc, Teodora Isoc, Technical University of Cluj-Napoca).

Second section "Modern Technologies" had one plenary lecture "Optimization of the Manufacturing Cycles by Means of the Continuous Improvement" (Constantin Bungău, Traian Buidos, Mihai-Dan Groza, Mircea-Petru Ursu, University of Oradea) and 8 oral presentations: modernization of a machine tool, vertical axis wind turbine and pale with variable profile, controlling the cooling rate of the bronze bell casting by adjustable coating, fast prototyping in product design, the study of the thermal problemes in ZIT, experimental determination of dilatation and contraction caused by the local heating of oxy-acetylene flame, experimental determination of the bending of thin plate in free state caused by the application of welding rows, magnetic control in welding.

Section 3 "Automation and Electrical Engineering" included 9 oral presentations where scientists discussed the next problems: CMOS image sensors and the perspective of the sequential cumulative exposure, artificial neural network for electromyographic classification, using fuzzy logic for a production line investment decision. case study, the transient operating conditions of the dual-windings stator induction generator, on watergy and aquifers, the romanian energy market.

Section 4 "Theoretical Mechanics, Strength of Materials, and Machine Parts" included information about study of the execution elements with piston. modeling and simulation, effect of shear deformations on bending beams, investigation of the forced gearing.

Section 5 "Mechatronics, Precision Mechanics and Micro-Mechanics" included 7 oral presentations: MEMS-based detection of the attachment of small colonies of cells, optical chopper wheels: a manufacturing technology, experimental setup for the study of optical modulators with rotating wheels, contributions to the analysis and design of polygon scanners, recent developments in aircraft wireless networks, fractional charges in advanced materials.

Section 6 "Education and Innovation; Railway Vehicles and Transport Systems" had 5 oral presentations: the triple helix model in innovation, considerańii privind invănămantul superior tehnic din Romania, on the shock loads that appear during buffing of freight wagons, improvements of monitor system of air quality in the Arad area.

Session "Education Sciences" had 4 invited presentations 3 sections: "Postmodern values and problems oral communications", "Curricular reform and school development in the 21st centuryoral communications", "Paradigms of social work in postmodern society oral communications".

The next 4 sessions were outside of interests of contributor of this paper and information about its are not included in this article. It was also some problems to visit all sessions because all of them were in parallel.

The next fourth similar symposium will be in two years in AVU (Arad).

INDEX

C

N

O

P

Q

T